Surfing the Solar System

by
James J. Wood, Sr.

authorHOUSE™

1663 LIBERTY DRIVE, SUITE 200
BLOOMINGTON, INDIANA 47403
(800) 839-8640
WWW.AUTHORHOUSE.COM

First published by AuthorHouse 05/03/05

ISBN: 1-4208-4452-0 (sc)

Library of Congress Control Number: 2005903145

Printed in the United States of America
Bloomington, Indiana

This book is printed on acid-free paper.

Dedication

Through out the years this data was being learned and manipulated I was married and had two children. The sweetheart of my youth passed away on April 11, 1997, after we had enjoyed forty-five years of Marriage.

I dedicate this book to my wife, E Jean Peterson Wood, and our two children, James John Wood, Jr. and Cecelia Elizabeth Wood. Without their encouragement and understanding this book and my many years accumulation of notes and calculator tapes would not have been possible.

At the end while the scraps were reviewed and the effort was made to put the stuff together in some form that might qualify as a book I was lucky, again, to have a companion willing to leave me alone to work when work was called for and to be there for me when help was needed. I must thank my special companion and friend for being with me through out the final workup. Thank you, Mrs. Vera M. Newman.

These comments seem to me now to be a very small offering for the love I received.

Mr. James J. Wood, Sr.

Table of Contents

Chapter One   An Introduction 1

Chapter Two   An overview 8

Chapter Three   My search for alternative Mass 17

Chapter Four   Planetary Mass in Miles 30

Chapter Five   Planetary Mass in Miles 39

Chapter Six   Planetary Mass in Miles.................................... 48

Chapter Seven   Planet Position Numbers 58

Chapter Eight   Satellite position Numbers 67

Chapter Nine   Saturn's Satellite Numbers 77

Chapter Ten   Uranus, Neptune & Pluto 83

Chapter Eleven   Developing Details 95

Chapter Twelve   A Little Digression 107

Chapter Thirteen   The Speed of Light 118

Chapter Fourteen   The Oblate Sun.................................... 129

Chapter Fifteen   Sun's Differential Rotation........................ 136

Chapter Sixteen   Let's Go to Mass
And a little on Surface Gravity 143

Chapter Seventeen   A solution to Rotation 157

Chapter Eighteen   Conclusion.. 173

Acknowledgements

I have not tried to research when the very first knowledge of Star watching is recorded but we can surmise it was a very, very long time ago. The Egyptians were hard at it over 3,000 years ago. Every major civilization has a history of astronomical interest of some kind. The keeping of records regarding the stars and planets seems universal. I have never personally observed or recorded any heavenly activity. My source of information originates from the efforts of astronomers and other academics. I have the utmost respect for all scientists of every pursuit. I had to lean upon the work and discoveries of others just to get started and then many times over and over to compare my results with those developed by academics the hard way with years of patient observations. Otherwise the three big names, to me, of discovery in this Solar System are Nicolaus Copernicus (1473-1543) with his conclusion that the sun is at the center of the Solar System; Johannes Kepler (1571-1630) with his discoveries on the functions of elliptic orbits; and Sir Isaac Newton (1643-1727) for his Law of Universal Gravitation. There are such an endless number of past and present big names in astronomy and related fields that I would not even venture to list any of them. I do not suppose to follow in kind by any means but I will be sharing my own ideas on the subject.

From the beginning of my efforts to discover curious as well as meaningful things about this Solar System and its various parts I have relied on the published works of many Astronomers and other learned and talented academics. It takes an expert to provide the details of planets just by looking at them from Earth. They give all kinds of statistics for the members of this Solar System and, needless to say, a guy like me would be in deep space trying to discern even the simplest data with my back yard binoculars. That was never part of my efforts. I had no serious interest in stargazing and I leave it to the experts.

My one observation that I can not leave unmentioned is that the differences reported in various text books and other source material relating to specific characteristics of the planets, such as speed of rotation and the like, caused me some repeated efforts that have still not resolved. Some conflict is to be expected; I see that now, but during the course of working with the various reported figures over and over it was frustrating. For this reason I will disclose or credit a source for data as I go along throughout the book.

Justified or not my intuition has caused me to formulate certain pre-conceptions about this Solar System. This book deals with the extremely curious numbers within the system that in turn arose from my efforts to discover reasons for some basic phenomenon that may or may not be explained herein.

They are:

1. The cause for the orbiting velocity and rotational speeds of the planets.
2. The cause for the speed of light and possible unrecognized unique characteristics.
3. Some miscellaneous mathematics intended to show the simplistic side of the System.

Chapter One
An Introduction

We have inherited a lot of information from our predecessors. Many authors speculate that there were ancient civilizations before recorded history. Our civilization may have lost more knowledge from the past than it has inherited. It is fun to speculate about such things like a lost continent of Atlantis, or to conjecture about flying machines as recited in the Indian literature. The possibility of past visits to Earth by extraterrestrials is often argued but never proven by any objective evidence. We may speculate that some artifacts seem inexplicable or survive without adequate explanation as to the source but that does not persuade the skeptic. I have opinions of my own but they are simply opinions that do not serve to resolve any of the issues. I think of the past in terms of simple mathematics and then wonder at the source. The Solar System appeared to be fertile ground for my efforts.

We have inherited a system of measures for distance and time. They gave us miles, yards and feet for distance and years, days, hours, minutes and seconds for time. When I talk about the Solar System I use all of these measures to describe how I think the system works. Astronomers like to use kilometers, meters and centimeters instead of miles, feet and inches. For larger distances they use the 'astronomical unit', which is the mean distance of the Earth from the Sun measured in kilometers. According to

1

one source this is about 149,600,000 kilometers (* .6124 = 92,961,440 miles). This is shown as 1 AU referred to as one astronomical unit. The 'mean' measure is used here, as it will be, for all planets and moons, because the orbits of these bodies are not true circles. They are oval, elliptic. Even so the planets vital statistics will measure out as if the orbits are circles for all practical purposes. We can thank Johannes Kepler for that result. I do not make much use of AU measurements. When you measure the various circumstances in terms of an AU yardstick you get Earth related data for results. I do not use Kilometers, meters or centimeters to demonstrate any criteria—unless I am quoting a formula of some other person's creation that uses such measures. The next most often used measure by professionals is grams and kilograms (kg) for the purpose of designating weight and Mass. The use of "mass" with regard to a planet or to a satellite is a means of describing what the object is believed to weigh. When we get to a discussion of Sir Isaac Newton's Universal Law of Gravitation we will be dealing with his discovery of a formula to measure the relative weight of the planetary bodies. No one can question the accuracy of his method for what we see now in the Solar System. I do wonder if there is some relative mass factor dependent on a planet's position relative to the elliptic, planetary tilt and the like. I will provide some amateurs thoughts on this point.

A little data about me would appear to be called for lest you think I have some special training in the field of astronomy-I have none. I can sum up my life as a high school drop out that found out later that was not a smart thing to be. After some time in the Army I met my sweetheart and we were married in California where we settled down and had two children. I was born in Philadelphia, Pennsylvania. My wife was born in California so it was not too difficult to decide to live in California. I went back to school and was sworn in as an Attorney at Law and was a very busy trial lawyer specialized in administrative law until I retired in 1994. I have always been interested in mysterious subjects. It was part of my interest in the large pyramids of Egypt and the unusual dialogue of the Bible that made me switch my interest to the Solar System. Looking back I would be hard pressed to explain how, or why, that transition took place but it did. Once I started to read the astronomical stuff I could see that there were still some areas where in an amateur could look for answers. My efforts over these many years were in the evenings and on weekends when there was no family activity scheduled. We still had time together for fun, trips or to sail our sloop.

One area dealt with the rotation of the planets. No one ever figured out a mathematical formula for that. Over the years the academics kept changing their data as the space ships got closer views of the more distant planets like Uranus and Neptune for example. I would knock myself out working on a formula only to find there were new figures to resolve. That effort is still in the works. I have been very close but I'm still not sure of the final answer.

Where to begin? Possibly the suggestion, without any objective support, that on some very distant past occasion some very smart people devised a system of measures that not only served to describe their celestial surroundings but also provided the means by which a simple exploration would be possible. To accomplish this goal the measures must intertwine between distance and time factors. We know it is not possible to describe the orbital velocity of a planet without the input of the distance covered in some related time frame. Any use of time must be specific because we are dealing with small numbers compared to large distances. At one point while preparing my outline for this book I thought a full research of the origins of our miles and time references would be important. After starting such research I could see that it was going to be a wasted effort. There is no exact origin available that I could find. Someone can argue that the Babylonians were the first to use the 60 measures that we get in hours divided by minutes divided by seconds. Does that prove they were the first? Would it even be credible evidence as to where the method originated? No. So I didn't need the extra work. Some offer that the 360-degree circle, the foot and its 12 inches may be from the Babylonians as well. This suggestion of the origins looks reasonable to me and better than I would have to offer. The accuracy of this view is not important to this book. We have the measures and that will work for us.

Astronomers accepted the diameter of the Sun at 864,000 miles for many years. More recently they talk of layers and such which may alter their view somewhat but I use 864,000 miles as the fact. Now we must pause here a moment because this may be the most important conclusion we will make. The measure could have been 50,000 drones or 400,000 clogs all of which would have meant something to people creating the measurement system. That would be no different from saying that the Suns diameter is 1,390,408 kilometers. If we assume that the measure of the Sun will remain the same, ('a rose is a rose.') no matter how we name it, then our secondary area of concern should be how our system of measures will serve us for the rest of our Solar System. Our choice, of necessity, must take the flexibility of time into account. Some very clever unknowns discovered that there was a way to make it work. Take the home planet, such as an Earth day, and divide it as a part of the Sun's measure for distance so you have a symbiotic time/distance relationship.

The Earth has a time schedule providing 86,400 seconds in a day. This is the generally accepted value. On Earth we accept 24 hours as one day, which means that the Earth makes one full rotation of 360 degrees in 24 hours. As 24 * 60 * 60 = 86,400 seconds. Remember that all references herein will be expressed in miles with 5,280 feet being equal to 1 mile.

The Sun with a diameter of 864,000 miles suggest a curious relationship of numbers. There are many numerical curiosities between the Sun and Earth. The Earths orbital velocity at 18.505 mps^2 x Earths diameter of 7,926.6 miles = 2,714,345.4 and that is the Suns circumference in miles. Earths seconds in a year, (86400 * 365.25) = 31,557,600 / 2,714,342.4 = 11.626228. This result multiplied by 3.1416 = 36.525 that is Earths days in a year divided by 10. Working only with miles let's try 864,000*3.1416= 2,714,342.4 = Sun's circumference and that divided by the equatorial diameter of the Earth of 7926.6 miles will = 342.43. The square root of 342.43 is 18.505 and that just happens to be the mean orbital velocity, in miles per second, of the

James J. Wood, Sr.

Earths speed of revolution around the Sun.

It is very important to note that if you used kilometers for this calculation you would end up with 18.505 as above. However 18.505 kilometers is a useless and meaningless result and a waste of time.

This example is offered to show we are dealing with exact numerical relationships that defy any suggestion that these results are merely co-incidental.

I expect that the reader will enjoy this book more if there is a sense of mystery included along with all the numerical renditions. Some of my suggestions will come as a shock to many academics. The speed of light has been sanctified without any offer ever having been made as to why it has a fixed speed. Light has been measured here on Earth at about 186,281 miles per second. Why 186,281 miles per second? Why would it always stay at that speed? Maybe I can help with that issue.

Planets revolve around the Sun in specific orbits based on the theory that they are "falling into the Sun" but retain there respective positions due to their motion, which being uninterrupted, will carry them on their repetitive way forever. Why? It is well established that the orbital velocities of the planets decrease as you get farther and farther from the Sun. I think they have figured that result is due to less attraction of gravity as you get farther from the sun.

There will be some repetition to deal with because this book is based upon hundreds of notes and calculator tapes that

I had to reacquaint myself with to extract my previously worked out calculations. A little repetition is not all bad.

One last point is of importance. I want to stress that there is no intent to challenge the establishment discoveries and theories. I recognize my limited abilities and I have no special program for a confrontation. My views are nothing more than a statement of what I think about the Solar System and some ideas on how it works. I am either right, partially right or I am wrong as to every point I offer. I invite any person with an opinion on the subject to have a field day correcting any errors anyone discovers.

Chapter Two
An overview

The fact that the Solar System has some oddballs requires an explanation. To approach the problem in this Chapter I will isolate some of the wrongdoer's.

An ideal Solar System, in my mind, would have Planet's with circular orbits on the Sun's ecliptic plane with their axis perpendicular to the ecliptic plane. All Planets and moons would rotate counter clockwise in this system or in the same direction as the Sun rotates. All planets would revolve around the Sun counterclockwise in this System or in the same direction as their Sun Rotates. All moons would rotate and revolve about their parent Planets counterclockwise in this system. We have some of the ideal criteria in our Solar System but we also have some notable exceptions. Let's start from the Sun going out.

Before I did any reading on the subject I was of the mistaken view that the mass of the various planets played some part in the position of the planets orbit. Not so. There may be an ingredient of mass involved but it is not clear. If mass played a part I suppose that it was when the substance was forming and before it took shape. If the mass of a planet played a part in its position in the Solar System the most massive with the assumed greatest attractive gravity would reasonably support orbits farther away from the Sun than we see. Why is the orbit of the most massive planet, Jupiter, almost midway in the solar layout? The system, as I see it, if based upon the relative mass of each planet, should reasonably start from the Sun out with the smallest bodies progressing to the largest body at the perimeter. We do not see that here. An explanation is offered for us by the conclusions that, in a vacuum, all bodies "fall" at the same rate without regard to size or mass. The belief that the planets are "falling" into the Sun provides that the specific location of any planet in this system is random because the mass of the planet does not control its present location.

The elliptic plane is the plane of the Earths orbit around the Sun. This can be thought of as a big invisible flat plate. All other planets are measured against this hypothetical surface for indications of inclinations, or deviations, from the Earths elliptic plane. If we were to take the equator of the Sun as perpendicular to the Suns axis then the plane of the Suns equator should set the measure from which all deviations would be calculated. We do that with the Earths elliptic plane and as a practical matter the deviations are about the same.

I read that the planet Mercury has an advance in its orbit around the Sun by 43 arc seconds per century, more than they feel would otherwise be expected, due to the effects of the Sun's gravitational warping of space. This is getting over my head. This was one of the criteria used to confirm Einstein's General Theory of Relativity because the theory almost predicted this result. The conflicting view claimed that the Sun was oblate, making the Sun or a planet larger at the equator normally caused by rapid rotation of the body, and possibly the gravitational tugging of planets/ satellites and that the oblate ness was an explanation for the wanderings of Mercury. Just about every major planet in this System is oblate to some degree due to spin. If the Sun was sufficiently oblate it could account for the unique orbital activity of Mercury. This conflict was resolved in favor of Einstein's suggestion. The mainstream academics concluded that the sun was not sufficiently oblate to account for the orbital mechanics of Mercury. For a layman like me that still seems to be an interesting debate. The measurement of the Suns equator to determine if there is some degree of oblateness may not be visualized. This will be discussed as part of my various speculations on the speed of light, the question of differential rotation of the Sun and related academic concepts.

The planet Venus is a real challenge. Venus revolves around the sun in a proper counterclockwise direction but it rotates clockwise which is directly opposite to the mainstream protocol. This is described as retrograde rotation. The planet is boiling hot. This is thought to be due to the "greenhouse effect". Sunlight passes through the atmosphere to the surface heating it. Then, due to the environmental factors, the heat is trapped beneath the cloud layer. How did all of this come about? No one knows. The suggestion of a greenhouse effect is not an answer but rather a conclusion. Are we to believe that Venus was created with its present perversity? I think not. Venus is a ripe subject for speculation. NASA scientists have suggested that Venus, in the past, had loads of water, potentially an amount greater than what the Earth enjoys. I am inclined to believe that Venus was originally a larger planet with abundant water and located beyond Earth going towards Jupiter. How and why was it compressed to get so hot? Any ideas?

The Earth seems to be unique. We have abundant water and abundant life forms. The Earth does tilt about 23 degrees off of the perpendicular mentioned earlier and there must be a reason for this. The Earth also has a Moon and that is something that Mercury and Venus do not enjoy. Also our Moon is the largest such satellite relative to the orbited planet than any other satellite is relative to the parent planet. There continues to be some academic disagreement as to the origin of our Moon and this still persists. The Earth/Sun numerical relationship is truly unique from all of the other planets. That could in part be the result of the numerical system being slanted that way.

The planet Mars seems normal enough but it is tilted at an angle of 25 degrees. It is smaller than Earth or Venus. Mars has two small satellites, moons, orbiting its surface. Phobos and Deimos. The revolution period of Deimos has been noted to be other than would be anticipated based on gravitational considerations. There is no satisfactory explanation for this curious behavior but it has been speculated Deimos might be hollow or of very little mass as the cause. A potential factor may be reflected in orbital velocity being involved with relative mass. I am inclined to the view that Mars was originally located about where Venus is now but I cannot devise a means to prove it.

The Asteroids, primarily found in the so-called Asteroid belt, between Mars and Jupiter, are very interesting. These comparatively small objects that range from small rocks to objects some hundred or more miles across are numerous. Thousands are known. There are important questions to be answered as to their origin and as to their behavior. It is suggested that the destruction of what was once a planet in that area would not provide the answer because we are told that all of the Asteroids put together are not believed to be sufficient to make an object with the mass of our Moon. The most significant facts about these chunks are that they are not spherical and that they are relatively condensed in their orbits. I recall once trying to run some of the asteroid orbital velocities for the heck of it and there were some surprises. If it develops that there is something meaningful I will share it. The creation of the System favors spheres. Rocks are not spheres and for this reason I am satisfied that the rocks of this System were caused by collisions and other disruptive factors that followed the original creative process.

Jupiter, the largest of the planets, is tilted only 3.13 degrees and its orbital inclination is about 1.3 degrees. It is considered to be mostly gases, some of which may be metallic in its core. I suspect this to be a planet that probably remains in the same position and condition since it was born because it is so massive that we know of nothing that is likely to compete with it or move it. This is the fifth planet from the Sun. Why is Jupiter, our largest known planet, located where it is in the Solar System? The rest of the planets going away from the Sun get smaller.

Saturn, a beauty to behold, is the second largest planet in the system. The rings are not unique because other large planets have rings but they are simply not as visible because they are not as dense as Saturn's and not as large. Saturn's axis is tilted 25.33 degrees and its orbital inclination is about 2.5 degrees. This is the sixth planet from the Sun. Saturn was of particular interest to me because it displays a significant tilt that could possible give me some assistance in developing my idea about the apparent mass of a planet being effected by such tilt.

Uranus is a very interesting study. A large planet by terrestrial standards but somewhat disappointing when compared to its neighbors. Uranus is larger than its neighbor Neptune but has been measured as less massive than Neptune. That is fascinating to me. The planet is tilted about 98 degrees which means it's almost on its side. The orbital inclination is less than 1 degree. This is the seventh planet from the Sun with its major satellites in retrograde rotation relative to Uranus. Why would any planet appear to weigh less even though it is larger than its neighbor? We could argue that the location where Uranus was formed there was lighter stuff accumulated, OK, why was that? How about some uniform creative process that compacts lighter stuff into smaller packages- that seems logical, doesn't it? If so, why isn't Uranus smaller as well as lighter? Also note that all four of Uranus major satellites are in retrograde rotation, as is Uranus, and that adds to the mystery.

Neptune is the farthest of the large gas planets. Neptune is tilted 28.3 degrees on its axis and has an orbital inclination of about 1.8 degrees. It is decided that Neptune is more massive (heavier) then Uranus therefore it follows that it is believed that Neptune has more solids than Uranus in a smaller volume. All things being relatively the same in space why would one sphere collect itself in a smaller denser volume than another sphere? If Neptune is more massive than Uranus then there must be a discoverable reason. My first impression is that the means of measurement may provide an answer, then again, maybe not.

Pluto, the outsider is a dark object getting very little light from the Sun. For quite a while there was very little accurate data concerning Pluto. More recently I came across some information that the authors felt was reasonably correct. Pluto has an unusual orbit that is not at all comparable to the other planets except that it is elliptical. Pluto is believed to be tilted 122 degrees on its axis and has an orbital inclination of about 17.15 degrees. Pluto has a moon, Charon. Pluto is thought to be about 1,413 miles in diameter and Charon about 728 miles in diameter. Pluto's mass weighs in at about 6.4×10^{-9} solar masses or about 7 times the mass of Charon and .002125 of Earth's mass. Pluto's mean orbital radius is about 3,674,661,328 miles. No wonder it has been so hard to measure its statistics.

Incidentally, if this data for Pluto and Charon is correct then this team is more unique than the Earth/moon team with Pluto being only 1.94 times the size of Charon while the Earth is about 3.75 times the size of our moon.

There are more than enough mysteries to go around. We have simply jumped over the unresolved Solar System issues to enjoy the speculations about deep space. I do not have the understanding that even a dabbler might have when it comes to the Universe so in the interest of restraint I will leave it alone.

Before I go any further with this subject I want to note that there is a fellow, Mr. W.C. Wright, which voices the opinion that "gravity is a push" and this is diametrically opposed to the main theory that gravity is an 'attractive' force. When I heard of his view I was very curious as to just what it was he was arguing. I read a little of the stuff he was quoted as saying and thereafter decided to spend the 20 bucks to get his books/pamphlets. Science says that the Sun and the planets (moons too) attract each other so that the force overcoming the Earth falling into the sun is the speed of revolution that the Earth has around the Sun. But for the effect of gravity the Earth would fly off into space. That makes sense. Mr. Wright argues that gravity is a mutual push causing the Earth, which might otherwise fall into the Sun, to instead keep in orbit around the Sun that is pushing the Earth away. That makes some sense. The problem I had with the "push" theory is that many objects "fall" to Earth instead of being pushed away into space. Science tells us that the effect of attraction between bodies in space can be used to discover unseen bodies just by the behavior of the primary visible body. I like to keep an open mind on things but as of now I couldn't make room for the push theory. I had already satisfied myself, by my own calculations that the revolution of the planets around the Sun was due to the effect of sunlight in conjunction with the Sun's gravitational effects on the planets. I admit that it has appeal when dealing only with planets revolving around the Sun or moons' revolving around a planet- this is because it is hard for me to understand how planets, that want to fall into something, continue to avoid that mutual attraction for thousands if not millions of years. Also I did not see where Mr. Wright offered a mathematical theory for his view that would supplant the proven effects of Newton's work.

This is a very short outline of the Solar System's main participants just to make sure we have some preliminary data to start with. I wanted to stress some of the inconsistencies so you will have some idea of where I am going with this effort. I will be hitting on a lot of old stuff. My primary goal remains the solution for planets rotations but I cannot resist poking around the place. Solutions to these varied inconsistencies may be beyond my ability to speculate. Even so, I am at liberty to offer the best I can muster and keep in mind that my imagination may be more suited to the task than are my skills.

As of this writing there are no further planets discovered in the Solar System. There has been speculation that a large object was to be found beyond Neptune based on calculations of the planets wobble indicating some massive interference. The original effort caused the discovery of Pluto. The calculations were not really correct and the planet was not the size sought but it was a success any way. My speculations envisioned a very large object somewhere out there as the cause for Jupiter being in its position with the balance of the planets getting smaller as you leave the sun going out of the System. We also have authoritive commentary to the effect that the edge of the Solar System has a lot of stuff roaming about. Meteors, planetoids, rocks and clouds of dust are all suspected.

Chapter Three
My search for alternative Mass

Isaac Newton has the monopoly on determinations of the mass of objects in the System and I was hoping to find an alternative determination of comparative mass. This is not a matter of trying to reinvent the wheel because my goal was that if Newton's work was to be used by me, assuming I knew how to do it, I would get the same results that every one else was getting. That would leave all of my questions unanswered. It took me some time to come up with a method that worked for this purpose. Later in this discussion I will be using what I refer to as the SS#. This is what I decided was the probable circumference of this Solar System providing for a mean radius of 31,830,914,184 miles. This number was derived from what I contend is the point of orbit around the Sun wherein a body revolving around the Sun will travel at 1 mile per second orbital velocity. I originally felt that the mass of the planets might be a factor in their periods of revolution. Sir Isaac Newton's formula takes the period of the planet into account. This caused me to wonder if there might be some other way to work out planetary ratios differently. I had studied the data sheets provided by the academics. The orbital velocities of the planets was of interest, and I thought, that it should have a relationship to the mass of the body orbited and with elaboration provide some answers to the rotation question. Orbital velocity of objects is not normally published in that manner. It is necessary to interpolate and calculate the velocity from the "period".

Eventually I decided that if the Earth mass was sufficient to propel the Moon at a specific velocity there must be some relationship between the Earths mass, the orbital radius in miles distance between the moon and the Earth and the speed at which the Moon traveled in it's orbit. There may be other more discrete factors contributing to the orbital velocity of a planet or satellite but this was a start. This approach, I thought, may provide a simple means of getting a mass ratio. I used the mean orbital radius as equivalent to semi major axis. They say that the mean orbital radius of the Moon is about 239,239 miles for a diameter of 478,478 miles. The diameter * 3.1416 provides a circumference of 1,503,186.48 miles. We are told that it takes the Moon 27.32 earth days to make one trip around its orbit. The orbital circumference of 1,503,186.48 miles divided by 27.32 earth days equals 55,021.467 miles a day. 55,021.467 miles a day divided by 86,400 seconds a day equals an orbital velocity of .63682253 miles per second. If the Earth is capable of orbiting the Moon at .63682253 miles per second what is the ratio, in miles, of the Earths mass for rotating a moon at 1 mile a second?

I obtained the result by the square of .63682253, which equals .40554. This number times the orbital radius of 239,239 miles equals 97,022 as the radius of the mass of the Earth expressed in miles. There remains to be decided if this result is the mass miles for only the Earth or is it the value of the Earth Moon combination. I will save that inquiry for later. Of course it took me quite a while to hit on this because I am not mathematically gifted. This method can work for all of the planets but it provides some curious differences when applied to planets like Jupiter and

Saturn, which have many moons. We will find that there is no consistent value between the large planet and their moons wherein each calculation for each moon provides a different result and even with that the expression does not follow any consistent pattern. This is one reason I must suspect the values and planet – moon figures in combination. In spite of these differences I will set forth the results with regard to each planet, discuss the results lightly and eventually seek to explain what we see by this method, including any patterns and conclusions.

My reason for using a velocity of 1 as the measure of the planets force is that the number 1 cannot be squared. This provides a uniform measure for comparing the planets vital statistics. Beyond a velocity of 1 mile per second the velocity will be a fraction of 1 and continue to diminish until some other solar bodies force interferes.

It seemed to me that this method could also be used to measure the Sun's gravitational force in miles. Using the earth as my tool I checked and note that we are told the Earth's mean orbital radius distance from the Sun is about 92,961,440 miles. This times 6.2832 provides for an orbit circumference of 584,095,319.808 miles. It takes us 365.25 days to make 1 trip around the Sun. 584,095,319.808 divided by 365.25 equals 1,599,165.83 miles a day. 1,599,165.83 divided by 86400 seconds equals 18.505 miles per second. Using the same method as above I square 18.505 for 342.43 times 92,961,440 getting 31,832,785,899 as the mean radius of an orbit around the Sun with a velocity of 1 mile per second.

To recap, the Sun's mass in miles radius is 31,832,785,899 = minimum Solar System radius. The Sun's mass in miles diameter is 63,665,571,798. The Sun's mass miles circumference is 200,011,760,360, which I round down to 200,000,000,000 miles. 200,000,000,000 divided by 6.2832 = 63,661,828,367. This means that this Solar System has a minimum diameter of 63,661,828,367 miles and a radius of 31,830,914,184. I have been using the circumference of the system, at 1 mpsv, as 200,000,000,000 miles. I refer to this as my SS#. There are many interesting relationships to be found in this determination. The system is so mechanical that you can take the seconds in a year for any planet and divide it into the SS# and you will get the orbital velocity, in miles per second squared, for that body. I will chart some of this in another chapter. When we use this on planets with many satellites to ascertain the planets mass miles value we may discover some new mysteries.

The Sun's mass miles radius (smmr) of 31,830,914,184 divided by Earths 97,022 provides that the Sun is 328,079 times as massive as the Earth. One textbook describes the Sun as 330,000 times as massive as the Earth and another source makes the Sun about 332,948 times as massive. I assume their result was secured using Newton's laws. 330,000 − 328,079 = about 1,921. 332,948 − 328,079 = 4,869 that is like saying there are almost 5000 Earth size planets difference. The difference is significant. This will be discussed later. It is possible that we are seeing the published mass measured for the Sun, planets and satellites together.

The next logical question is to learn how the Sun relates to this SS# of the force in miles. The Sun with a diameter of 864000 miles provides for a circumference of 2,714,342.4 miles. SS# at 200,000,000,000 / 2,714,342.4 = 73,682.67 the square root of it is 271.44. To present another way take 2,714,342.4 * 271.44^2 = 199,991,862070, or about the SS#.

A real numerical surprise was when .000000000031416 popped up by dividing the number 1 by the Solar System radius. Like 1/ (200,000,000,000/6.2832). This is the solar radius. The Earth's equatorial diameter is about 7926.6 miles depending on which text book you are looking at for the information. Using 73,682.67 / 18.5 (ovmps) = 3,981.8 close to Earths radius as 2 times 3,981.8 = 7,963.5. To enlarge on the use and accuracy of the solar mass miles circumference of 200,000,000,000 / 18.505^3 (which are 6336.76) will provide the seconds in one revolution of the Earth around the Sun, or 31,561,870.7, seconds. This last number divided by seconds in a day of 86,400 provides 365.30 or almost exactly the days for Earths year.

This is a good place to show the proof that this method works for all of the planets in the Solar System. I have included Pluto even though the available data is less certain.

The published data on the orbits and velocity of the 9 known planets are:

Planet	Orbit Radius in Km.	Earth Days in a year	OVKmS	OVMS
Mercury	57,900,000	87.96	47.87	29.74
Venus	108,200,000	224.68	35.02	21.76

Earth	149,600,000	365.25	29.78	18.5
Mars	227,900,000	686.95	24.12	14.99
Jupiter	778,300,000	4,337.00	13.05	8.11
Saturn	1,427,000,000	10,760.00	9.64	5.99
Uranus	2,871,000,000	30,700.00	6.80	4.22
Neptune	4,497,100,000	60,200.00	5.43	3.37
Pluto	5,900,000,000	90,700.00	8.68	2.94

As shown earlier the procedure is quite simple. I convert the reported semi-major axis from AU or kilometers to get it in miles. My SSR of 31,830,914,184 divided by the mean orbit radius in miles gives me the orbital velocity of the planet ($OVMS^2$) squared with the square root being the actual miles per second velocity (OVMPS). When you compare the results in the following columns to the data above you will find the OVMPS are virtually identical. Only measurements in miles are used. This next chart is Sr#/orbit radius $=ovmps^2$ unsquared = mps velocity.

Planet	Orbit radius	Years days	Sr#/or $=mps^2$	Actual ovmps
Mercury	35,977,324	87.96	884.75	29.74
Venus	67,232,236	224.68	473.44	21.76
Earth	92,961,440	365.25	342.41	18.50

Mars	141,610,227	686.95	224.78	14.99
Jupiter	483,612,285	4,337.00	65.82	8.11
Saturn	886,695,015	10,760.00	35.89	5.99
Uranus	783,953,320	30,700.00	17.84	4.22
Neptune	2,794,363,106	60,200.00	11.3	3.37
Pluto	3,666,222,305	90,700.00	8.68	2.94

This works for the planets but it would also work for any object in orbit around the Sun. The same procedure will be used for the planet-moon relationships.

You will note I quote the "days in a year" from the textbooks. This is deceiving because they mean the number of Earth days in the other planets year- not the actual number of rotations the other planets will make in one revolution around the Sun. This approach had to be discarded if I was to make any progress on the formula for planetary rotations. The larger planets move in much larger orbits and compound the difference by moving at much slower velocities. All taken from published data.

	Orbit radius	Sidereal days	Rotation seconds	Estimated rotations
Mercury	35,977,324	87.96	58.65 d	1.5
Venus	67,232,236	224.68	243.01 d	0.924

Earth	**92,956,955**	**365.25**	**86,400.00 s**	**365.4**
Mars	**141,610,227**	**686.95**	**88,642.44 s**	**669.62**
Jupiter	**483,612,285**	**4,337.00**	**35,427.60 s**	**10,575.85**
Saturn	**886,695,015**	**10,760.00**	**36,838.80 s**	**25,247.76**
Uranus	**1,783,953,320**	**30,700.00**	**55,800.00 s**	**47,601.18**
Neptune	**2,794,363,106**	**60,200.00**	**56,880.00 s**	**91,324.52**
Pluto	**3,666,222,305**	**90,700.00**	**6.3874 d**	**14,197.58**

The goal here is to calculate the actual rotations a planet will have as it makes one full revolution around the Sun. Earth days are not a factor in the description. Using the actual rotations of the planets is important to seeking the formula, if there is one, for why and how the planet rotations come about. A further factor in the planetary rotations is that the larger a planet is the faster it will rotate. This is consistent for all of the planets. It is well worth noting that Uranus has fewer seconds in its day than Neptune meaning that it rotates faster than Neptune even though Neptune is said to be more massive than Uranus albeit smaller than Uranus. I will be demonstrating in follow up chapters that the Solar System is as specific as to velocities that it is almost as if there were circular tracks running through the System waiting to accept any revolving body that is engaged.

The calculations for the approximate rotations recited in the last column above are very simple. I am using the academics observed rotations for each of the planets. I start by converting the mean orbital radius into an orbit circumference in miles. The circumference is divided by the orbital velocity in miles per second providing the seconds in a full orbit. The total seconds of one complete orbit are then divided by the seconds in one complete rotation. The result should be fairly accurate for the total rotations in one complete revolution.

There is no single set of numbers to permit the assertion that my projections are the best they could be. From my standpoint the Earths mean orbital radius seems best set at 92,961,440 miles and I have preferred to use that in many of my original calculations. The use of the planet rotations as part of further work on the Solar System picks up in chapter five however this appears to be a good place to discuss what I call the invisible tracks throughout the Solar System.

Note that for each and every mile that you get away from the surface of the Sun there is a measurable difference in velocity for any object occupying that position. When dealing with planets that cover thousands of miles of potential tracks you will get a specific potential velocity for the planet edge closest to the Sun and a different (slower) velocity potential for the planets edge farthest from the Sun.

VIEW FROM TOP

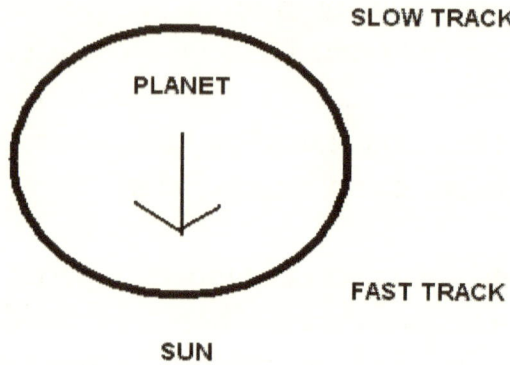

Using the Earth as an example the upper line will represent the outer edge of the Earths equatorial surface and the lower line represents the Earths edge closest to the Sun. Using my rounded SS# as the radius of the Solar System where objects would revolve at one mile per second I have 31,830,914,184. We know that dividing that number by the planets mean orbital radius will give us the planets orbital velocity squared. I have assumed for this purpose that the mean orbital radius generally refers to the center of the planet. We can recognize that if instead of the Earth there was a small planet one mile in diameter at the farthest location and a similar tiny planet at the inner location they would be, based on Earths diameter, about 7,924 miles apart. They would each have a specific orbital velocity based on their orbital position with the farthest the slowest. When we have one large planet covering the same area the situation is unchanged as far as the exertion of forces on the planet are concerned but because the planet is a solid object there is but one velocity controlling the planet. I assume for now that the force is acting on what would be the center point of the planet. I will demonstrate what I mean the long-winded way. Recall SS# is 31,830,914,184.

(SS#/(92961440+3963))= 18.5039257773^2 slowest tract.
(SS#/ 92961440)   = 18.50432018963^2 actual velocity.
(SS#/(92961440-3963)) = 18.50471462718^2 fastest track.

I deduct slowest from the fastest to get a position number,

> 18.504714627182
> 18.503925777300
> #. 000788849882   This is Earth's position number.

1/. 000789=1267.349, this is at times easier to work with.

I will be providing a position number for every planet in another chapter. Position numbers can also be assigned for satellites revolving around planets. Hopefully all of this will be put to some worthwhile use. For the Earth, about which I have done a lot of experimenting the position number seems to have interesting comparisons. The Earths equatorial surface rotates at about .2882 miles per second. Otherwise you could say that the equatorial surface takes about 3.4698 seconds to rotate one mile. 1/.2882 = 3.4698. Using our position number for the Earth we will be able to work out some interesting results. 1,267.349 / 365.25 days = 3.4698 and that are the seconds per 1 mile of Earth's equatorial rotation.

Now I have a method for determining the mass of the objects in the System in terms of miles that I will call mass miles or MM. That still leaves open the question of whether I am getting the mass of the individual object or the collective mass of the object and all of its friends. My MM for the Sun is so much less than Newton's method provides I suspect there is another mystery not yet resolved. You will note I only used Earth data to get my MM for the Sun. I could take each planet in turn and multiply the mean orbital radius by the square of the orbital velocity in miles per second to ascertain what degree of variance will be disclosed in the final individual results. I have not done that ritual for the Sun-planet relationships but I know when I did it for the planet-satellite units there is a variance. As of this point I am uncertain just what the figures are telling me, I'll keep listening.

This last chart is provided to show the individual planet calculations used to arrive at my determination of the Solar System minimum radius in miles. The results are not uniform and there appears to be no pattern involved. I simply averaged the results to obtain a radius and converted it to a circumference and then settled on the 200,000,000,000 miles.

The method to obtain the sun's mass miles figure based on each planets orbit provides the following:

Mercury $35,979,060 * 29.74^2 =$ **31,822,312,848**

Venus $67,235,480 * 21.76^2 =$ **31,835,838,414**

Earth	$92,961,448 * 18.50^2 =$	**31,816.055,578**
Mars	$141,617,060 * 14.99^2 =$	**31,821,367,543**
Jupiter	$483,822,040 * 8.11^2 =$	**31,821,991,597**
Saturn	$886,737,800 * 5.99^2 =$	**31,816,240,937**
Uranus	$1,784,039,400 * 4.22^2 =$	**31,770,887,250**
Neptune	$2,794,497,940 * 3.37^2 =$	**31,736,833,654**
Pluto	$3,674,648,900 * 2.94^2 =$	**31,762,195,232**

Total of 286,203,723,053 / 9 = 31,800,413,672

Using the average of the above resulted in a unique number that I could never have envisioned. It provides for a Solar System circumference that I have rounded out to be 200,000,000,000.00 miles. As numbers go it's attractive.

Chapter Four
Planetary Mass in Miles

My Solar System minimum circumference for my convenience is 200,000,000,000 miles for a radius of 31,830,914,183 miles. This is my miles mass number, Sn. I have shown how accurate it is for finding the mean orbital velocities of the planets.

I decided it would be a good way to obtain the relative mass of the planets by using the orbiting satellites. At this point we must recognize that the nature of a Sun Planet relationship is inherently different from a Planet Moon relationship. The Sun is emitting energy in the form of light and solar wind and who knows what else. By comparison a planet is inert and only has its gravitational mass to maintain the motion of a moon. Whether the planet is assisted in this work by the Sun has yet to be demonstrated.

I want to use the same basic formula with the planets as the central body around which the moons are revolving. Newton worked this out a long time ago but his method is different and not as simple minded as mine so we will not need exponential numbers. Keep in mind that at this point, with my method, we are not measuring specific forces, or any supposed weight, but rather the apparent ability of a planet to cause a moon to revolve around it at an orbital velocity of 1 mile per second. The result is derived from the proven documented ability of a planet to cause a moon to revolve around it at more or less than 1 mile per-second.

By doing this we can put all of the planets on an equal footing for more detailed comparisons. The mass recovered is expressed in miles mass as opposed to weight mass in kilograms or tons.

The planets Mercury and Venus have no moons so we must deal with them later.

The Earth has one Moon that travels around the Earth in a mean orbital radius of about 239,239.00 miles. It takes the Moon 27.32 days to make one complete orbit of the Earth. This gives the Moon an orbit circumference of 1,503,186.48 miles that when divided by 27.32 days shows that the Moon travels 55,021.46 miles a day. Using 86,400 seconds in a day provides for an orbital velocity of .63682 miles per second. We did this before.

Orbital velocity of .63682 miles squared equals .405543. That multiplied by 239,239 miles mean orbital radius provides 97,022 as the Earths mass miles number. 97,022 miles radius represents the point in space wherein an object will orbit the Earth at 1 mile per second. We do not take the mass or weight of the orbiting body into account in this calculation. Therefore I believe it is the demonstrated value for the earth with out the contribution of anything else. In theory the mass of the Moon may very well play a part in the original orbital velocity of the Moon's orbit but hopefully we have eliminated that factor converting it into 1 mile per second. Nothing can be claimed with certainty yet.

I was originally concerned that possibly I was simply determining where the Moon would have to be located for the Moon to travel at 1 mile per second orbital velocity. This could be a factor if the Moons position is due in part to the contribution of the Moon's mass to its orbit of 239,239 miles radius. This did not appear to be a factor in the Sun-planet systems as was shown by the universal accuracy of each planets orbit velocity with out regard to the planet's mass under circumstances wherein the mass of the planet's varied greatly. The conclusion on the issue must wait for the results obtained for the other planets.

MARS

Mars has 2 moons. They are very small moons. The published data when changed from kilometers is:

	Mean orbit radius miles	Earth days	Mean Orbit miles per sec.	Mass miles
Phobos	5,828.7	0.3189	1.32918	10,297
Deimos	14,602.9	1.262	.84149	10,340

I will replay the method step by step.

Phobos:
(5,828.7*6.2832)=36,622.8, orbit circumference.
1 day of 86,400 seconds * .3189 = 27,552.96 seconds. One orbit.
36,622.8 orbit miles / 27,552.96 = 1.329181 miles per second orbital velocity. 1.329181^2 = 1.7667 * 5,828.7 = 10,297.7 miles mass for Mars based on Phobos orbit.

Deimos:
(14,602.9 * 6.2832) = 91,752.94, orbit circumference.
1 day of 86,400 seconds * 1.262 = 109,036.8, seconds, one orbit.
91,752.94 orbit miles / 109,036.8 = .841486 miles per second orbital velocity. $.841486^2$ = .708098 * 14,602.9 = 10,340.2 Miles mass for Mars based on Deimos orbit.

This, by averaging the two results provides a miles mass for Mars at 10,319 miles mass. I ran the numbers a few time to be sure there were no errors. A very slight variance in any of the data has the prospect of making the results change, and at times, significantly. Calculating Deimos orbit provides for a slightly more massive Mars than does Phobos orbit. If my approach is correct then this difference, although very slight, must be accounted for and some explanation offered for it. One possibility is that the mass of Phobos plays a part in the result. The difference is 42.5 mass miles. The calculations do not include any mass values for these bodies so the most apparent source for the difference would be the contribution of the mass of Phobos added to the mass of Mars. There is also the prospect that any satellite in an orbit that has a larger radius will automatically give the parent planet a larger mass miles result. Then again the orbit data could be just a smidge off of perfect and we could easily get such a difference.

I think all differences from what we seek should be reserved for study as we progress. You will recall that I was concerned in the Earth – Moon relationship that I could simply be calculating an orbit for the Moons mass at 1 mile per second velocity and not the Earths actual mass miles. This same concern could apply to Mars except we can conclude that with two orbiting bodies of very different masses the results would not even be close if the mass of the objects was contributing to the results. Phobos is about 5.625 times the mass of Deimos according to the published figures. In this case, as with all planet-satellite efforts, I am confronted with the need to understand individual satellite contributions to the behavior of their neighbors. Logic suggests to me that the presence of Phobos in an internal orbit joins with Mar's mass to provide an extra push bringing about Deimos faster velocity.

For now I must assume that my source information is accurate otherwise I would be uncertain about the input in every calculation that I make. If we alter either the orbit radius or the orbit velocity we can change the result for Mars mass in miles. That does not appear to be justified so there must be some other reason for the difference. The method may not be as accurate for planet moon combinations as it is for Sun planet combinations but, for now, there is no reason to reach that conclusion.

My curiosity caused me to check to see if Deimos orbital velocity should be reduced or enhanced to get the same result I did for Phobos. The result is that an orbital velocity for Deimos must be .839754 mps to provide a Mars miles mass (10,298) equal to that provided by Phobos. At .839754 mps Deimos would have an orbit radius of about 10297.77 miles. The reported orbital velocity for Deimos of .841486 is therefore faster by .001732 miles per second bringing about the mass miles difference. That is not much of a difference. My present working analysis is that the speedup of Deimos orbit is due to the presence of Phobos in an interior orbit. Another likely contribution could be a kick by Jupiter. I may return to this question later on.

I note in passing, regarding the actual mass of Mars is said to be .1074 of the Earths. My miles mass for Mars of 10,298 / 97,022 for Earth = .1061. You may recall that my mass miles estimate for the sun was less than the published estimate. Now we find that this effort tends to show that Mars is .0013 less massive than Earth than was previously predicted by other methods. I keep working this issue because I think it is meaningful and I already know that when I run the same method on the other planets they too will show as less massive than predicted.

Rethinking Phobos vs. Deimos relationship with Mars it seems inescapable that Deimos is not much farther from Mars than one would expect. Phobos mass appears to be negligible in this regard. I mentioned previously that some academics have questioned the Deimos orbital properties as exceptional. I would be satisfied to leave it there but for the fact that the large planets will present some equally interesting results. I saved them for the next Chapter.

On the Internet I found a source for the stated values of the radius of the planets and some of the moons at **http:// enclyclopedia.thefreedictionary.com**. This source states that Phobos is 13.4 x 11.2 x 9.2 km. Deimos is stated to be 7.5 x 6.1 x 5.2 km. These rocks are not spherical and were almost certainly acquired by Mars some time after Mars was completely formed and most likely about the time the original planet between Mars and Jupiter was destroyed. I have another formula that I use to double check the accuracy of the data and I need the radius of the moon in order to apply it. Using this data I will assume a hypothetical diameter for Phobos at 17 miles and for Deimos at 7 miles.

Then, we can review the previous data:

	Mean orbit Miles radius	Earth days	Orbit Velocity miles per sec.	Mass Miles
Phobos	5,828.7	0.3189	1.32918	10,297
Deimos	14,602.9	1.262	.84149	10,340

Phobos,   8.5 estimated miles radius.
9380- 8.5=9,371.5 Mn/9,371.5=110110441231   1,04933522399
9380+8.5=9,388.5 Mn/9,388.5=1.09911061404 -.<u>1.04838476431</u>
                                                    #   .00095045968

$((8.5*6.2832)/\#) = 56{,}191. / 86400 = .65$ Earth days. This is twice as much as observed. This requires we do the effort in reverse to see what the equivalent radius of the moon is. Days of .3189 times 86400 equal 27,552.96. 27,552.96 * # = 26.19, 26.19 / 6.2832 = 4.17 miles.
Assuming the orbit data is correct Phobos may have an effective radius of 4.17 miles presented to Mars.

Deimos,   7.estimated miles radius.
23,500 -7. = 23,493 Mn/23,493=.043923721959 = .66274974129
23,500+7. = 23,507 Mn/23,507=.043897562428 =<u>. 66255235587</u>
                                                    # .00019738542
((7*6.2832)/#)= 222,825. 222,825 / 86,400= 2.58 Earth days.
Observed days are 1.262.
Now, Deimos in reverse to estimate a probable radius in miles.
Days of 1.262 times 86,400 = 109,037. 109,037 * # = 21.52. 21.52 /

6.2832 = 3.425 miles. Assuming the orbital data is correct Deimos may offer a radius of 3.425 miles facing Mars.

When the satellites of Jupiter are discussed this method is used for all of them and works extremely well. The underlying basis for this relationship between the orbit, the moons radius and the days is not yet clear to me. The orbital data is obvious because we feed that information into the original formula when arriving at the "#". The radius of the moon is used to extract the center point of the moon in relation ship to the orbit and may reflect the submission of a more accurate orbital result for determining the Earth equivalent days.

The above estimates of the hypothetical spherical radius of Phobos and for Deimos are most likely incorrect. Phobos at 13.4*11.2*9.2 Km's would have a volume of about 858 miles. Deimos at 7.5*6.1*5.2 Km's would have a volume of about 148 miles which makes it only .17th the size of Phobos. My guesswork provided a comparison with Deimos being about 82% the size of Phobos. I will try again using my estimated miles radius.

Deimos rework

((3.425*6.2832)/#)= 109,025.07.109,025.07/86,400 = 1.262 days, exactly the same. This could mean that the effective spherical, radius of Deimos is 3.425 miles as presented to Mars.

I will rework Phobos again to double check the result.

Phobos rework

((4.169*6.2832)/#) = 27,560   27,560/ 86,400 = .3189 days and the same as published days for Phobos orbit. There is a conflict here with the published data. Phobos is much larger in the published data than Deimos yet by the above approach they measure quite closely in their radius.

I do not have a means as yet for estimating the spherical value for rocks that are irregular and tumbling along with a different presentation to the planet and the orbit with each tumble or rotation. My use of the above formula for double-checking the orbit characteristics with the days in the orbit has been shown elsewhere to provide very useful results. All I can conclude from the above is that no matter what the volume of Phobos may be it presents an image to Mars and the sun that is consistent with a sphere radius of 4.169 miles.

I too, am simply tumbling along until the data in the puzzle presents a useful picture.

There is always reason to be cautious and make sure you are not feeding in the data that comes back to you as the result. In this instance we provide the mean orbital radius and the miles per second velocity to find the difference between the planets side nearest the sun and the planets side farthest from the sun. The subtraction and addition do insert the diameter of the planet into the formula when arriving at the planet or satellite significant number. I do not insert the days into the first work up unless that is automatic from the orbital velocity miles per second.

Chapter Five
Planetary Mass in Miles

You may find this discussion less than entertaining unless you are intrigued by the search for undisclosed ingredients that create the inner workings of our Solar system. To search and fail to discover is better than never searching at all. We left Mars with the knowledge that the mass miles disclosed for Mars was .0013 less than the published Earth-Mars mass ratio predicted for Mars. Our method of determination was productive to within 42.5 miles difference in the orbital characteristics between Phobos and Deimos. Jupiter is something else. The same formula will be used for all of the planets and satellites.

JUPITER

It is worth noting that there is no uniform satellite progression as to size of satellite and its location in orbit around Jupiter. An example is that Callisto is farther from Jupiter than Ganymede even though Ganymede is more massive. It may be helpful to include a chart showing the published mass for each satellite that is discussed in its proper order as you leave Jupiter. Later.

I found with Jupiter and the other large planets with many moons that the miles mass ratio for the planet varied with each different moon that was calculated. This could provide some interesting information. The mass of Jupiter is said to be 317.89 (or 318.35 by another source) times that of the Earth. If Earths mass miles are 97,022 (* 317.89) then Jupiter should have mass miles of about 30,842,323.58 miles. We can look at the moons ratios, as before, to check it out. I will dispense with showing the data in kilometers because it will save space. There is no moon that produces a mass miles figure over 30,775,003.56 and that particular result is well above the average.

Jupiter's Satellites		Orbit radius	Period days	OVMPS	Jupiter mass in miles
Amalthea		112,473.40	0.498	16.42	30,316,304.24
Io	L	262,230.80	1.769	10.78	30,474,013.02
Europa	L	416,959.40	3.551	8.54	30,402,846.28
Ganymede	L	664,898.00	7.155	6.76	30,365,471.87
Callisto	L	1,168,232.00	16.69	5.09	30,269,753.20
Leda		6,903,753.99	239.0	2.1	30,464,525.62
Himalia		7,146,099.98	250.6	2.07	30,731,192.44
Elara		7,270,379.98	259.7	2.04	30,134,341.65
Lysithea		7,394,659.98	263.6	2.04	30,775,003.56
Ananke		13,173,679.97	631.1[R]	1.52	30,357,028.81
Carme		14,043,639.97	692.5[R]	1.47	30,544,554.65
Pasiphae		14,602,899.97	738.9[R]	1.44	30,163,397.34

Sinope	**14,727,179.97**	**758.** [R]	**1.41**	**29,400,490.60**

The average Mass miles for Jupiter, totaling all results and / 12 = 30,416,536.05. The average Mass Miles for Jupiter, totaling 4 largest moons / 4 = 30,378,020.75.

A preliminary comparison of Jupiter / Earth shows 30,378,021 / 97022 = 313.1 times as massive as Earth.

This suggests that Jupiter may not be either 317.89 or 318.35 times as massive as the Earth.

I have alluded to a potential wave pattern in the numbers for the satellites. Here, we find that the first smaller orbit starts with a smaller mass miles only to jump at about 250,000 miles and continually drop thereafter until about 1.2 million miles. It then goes up and down every few million miles. A graph would be best to find a wave but this book is not large enough to show one. When I checked these results I was tempted to think the miles mass formula was not working. You will note that no two moons produce the same miles mass for Jupiter. This result is similar to that for Mars except that the Moons farthest from Jupiter are not uniformly showing more miles mass for Jupiter, as we will find for all of the major planets. I will assume that all of the input data is correct. There is no apparent relationship between the result and the size or mass of any particular moon. The four largest moons of Jupiter are Io, Europa, Ganymede and Callisto, in that order counting from Jupiter outbound.

Moon	Jupiter radius	Moon Mass Kg	Moon density	Mean MM
Io	30,474,013.02	1,128.46	8.916×10^{22}	3.55
Europa	30,402,846.28	971.24	4.873×10^{22}	3.04
Ganymede	30,365,471.87	1,639.25	1.490×10^{23}	1.93
Callisto	30,269,753.20	1,497.57	1.064×10^{23}	1.81

The progression is to provide less miles mass for Jupiter for each successive moon as you go away from Jupiter. However that does not hold true for Lysithea at 30,775,003.56 or for Carme at 30,544,554.65. Carme is in retrograde revolution so that may be a factor. Ganymede is the largest and most massive of the moons even though it is stated to be less dense than Io or Europa. This last sequence isolating the larger moons does have one item of note relating to the progressive decrease in density of each moon as you go away from Jupiter. The diminished density of each moon does coincide with a reduced miles mass result. The size of the moon fails to be suggestive and seems random. I keep looking for a pattern or a common denominator and in this sequence it appears that the moons density and /or the moons mass are the most interesting items for the sliding mass miles result. Unfortunately this is not uniformly true. I think I am getting results for the planets mass miles determinations that also reflect a contribution for the moon itself.

I decided to rework the calculations using a mass miles for Jupiter at 30,378,021. This is the result gained from the four largest moons of Jupiter in the last workup. In addition I decided to use a different source for the data as to the moons distance from Jupiter and as to the Sidereal period in days. This chart will show the moons orbital velocity as accurately calculated from the published periods in days and the calculated orbital velocities using Jupiter's tentative Miles Mass of 30,378,021.

The method is simple. The listed day's times 86400 seconds gives us the seconds in one revolution of the moon around the planet. The mean orbital radius times 6.2832 gives us the orbit circumference that is then divided by the seconds in one orbital revolution and that provides the mean orbital velocity in miles per second. We know from previous efforts that a miles mass number for Jupiter can be a common denominator for orbital velocity.

Jupiter revisited

Satellite	Mean Radius	Orbit Period days	Orbit vmps	Jupiter's MM=Jn	OVMPS using Jn
Amalthea	112,660	.498	16.45	30,486,077.	16.420
Io	261,982	1.769	10.77	30,382,240.	10.768
Europa	416,897	3.551	8.537	30,383,607.	8.536
Ganymede	664,898	7.155	6.757	30,357,280.	6.759
Callisto	1,170,096	16.689	5.098	30,410,331.	5.095
Leda	6,897,540	238.7	2.101	30,447,127	.2.098

Himalia	**7,127,458**	**250.6**	**2.068**	**30,481,457.**	**2.064**
Lysithea	**7,276,594**	**259.2**	**2.0415**	**30,326,822.**	**2.044**
Elara	**7,297,100**	**259.7**	**2.0433**	**30,465,938.**	**2.0403**
Ananke	**12,862,980**	**631.**	**1.482**	**28,251,271.**	**1.5367**
Ananke-	**13,173,680**	**631.1**	**1.518**	**30,356,426***	**1.5185**
Carme	**13,888,290**	**692.**	**1.459**	**29,563,739.**	**1.4789**
Carme-	**14,043,640**	**692.5**	**1.474**	**30,512,179***	**1.4707**
Pasiphae	**14,478,620**	**744.**	**1.415**	**28,989,434**	**1.4484**
Pasiphae-	**14,602,900**	**738.9**	**1.437**	**30,163,399***	**1.4423**
Sinope	**14,727,180**	**758.**	**1.412**	**29,362,226**	**1.4362**

It is quite obvious that the use of my tentative Jn for Jupiter, based only on the four largest moons, is accurate enough to produce almost identical orbital velocities as the true observed orbital velocities. There are some noteworthy differences between the content of the source material. The moons Lysithea and Elara are in reverse order when compared to the first chart. That is because Elara is now shown as being in a larger orbit than Lysithea but strangely enough the orbital velocity is still faster than for Elara over Lysithea, which is not the way it works normally. The larger the moon's orbit radius is from the planet the slower the orbital velocity will be normally. I have run the two moons data many times and always get the same result that causes me to believe that this source data is not correct for these two items.

For Ananke, Carme and Pasiphae I included the data from the previous chart for comparison with the current chart. In spite of the differences in the data the method has proven effective and accurate to a significant degree. This makes one think that the masses of the various moons have nothing to do with the planets orbital position or its period of revolution. These varied size moons show no pattern that could be related to mass. The relative mass is shown just for an idea of values. The fraction is too small to fit the chart. Later on a graph of the values based on each moons mile mass result may show a pattern in some wave that will help with all the large planets. As of today I do not have access to published data for the rotation of these satellites. As we progress something may come up to include more data for comparisons.

Jupiter is huge and a fascinating subject to contemplate. Jupiter's location in the Solar System is itself a mysterious circumstance as I see it. The big red spot always causes me to think of the Biblical reference to the alleged mark put on Cain 'by the Lord' after Cain slew Able. This comment has nothing to do with our subject in review but my thoughts do travel in strange ways at times. The likelihood of a missing planet between Mars and Jupiter bears a humorous connection to the Biblical story. It is an unusual marking that I believe is attributed to a storm in Jupiter's atmosphere.

In another chapter I deal with the rotation of the planets in some formulas I worked out. I have some space here so I will use it to seek the result for Jupiter's larger moons. First we must find what I call the position number.

You will recall we want the center point of the planet or moon, which we get by first deducting the objects radius from the mean orbital radius and then adding the objects radius. We run out the orbital velocities and deduct the slowest from the fastest.

Ganymede, Jn= 30,378,021, orbit radius = 664,898.
Jn / (664,898 − 1,639.25) = 45.8, unsquared = 6.7676555
Jn /(664,898 + 1,639.25) = 45.5, unsquared - 6.7509909
$$\#\quad .0166646$$

I do not know Ganymede's rotations. The sidereal period is shown as 7.155 Earth days. Orbit radius times 6.2832 provide 4,177,687 orbit circumference. Divided by orbit velocity of 6.759 will equal 618,092.47 seconds in a day divided by 86,400 equal 7.1538 days. I did all this to show the days work out as were recited. So at it's most basic the rotations will normally be (days * #), 7.155 * .0166646 = .11923 equatorial surface velocity in miles per second. Or, 1/.11923 = 8.386 seconds per mile of rotation. This will work properly only if the actual rotations of the planet are used and not the Earth's equivalent. We are using Earth seconds in a day here and not the actual seconds in one day on Ganymede.

I will jump ahead a little bit here to use another formula I developed to get accurate results for the equatorial surface velocity of rotation for orbiting objects.

1/(((Orbit radius/planet radius) / days) / ovmps) = evmps
(664,898 / 1,639.25) = 405.61. 405.61 / 7.155 = 56.69.

(1 / (56.69 / 6.759)) = .1192 miles per second equatorial surface velocity rotation for Ganymede.

Not knowing what the published data was for Ganymede I went to the JPL web page again. They do not provide the surface velocity rotation of objects. They show (deg/day). I interpret that to be 360 degree divided by 7.155 days. The result is 50.317 and that is what they show meaning the moon rotates 50,317 degrees in 7.155 days. It seems that 50.317 / 360 = .13976. That is the same as 1/7.155??
I am sure that I am doing something wrong here.
That measurement of the moons orbital rotational velocity does not do a thing for me. I do not find it meaningful.

Chapter Six
Planetary Mass in Miles

Now that we have made a preliminary exploration of Jupiter's major moons we should check to see what will develop for Saturn, Uranus and Neptune. I intend to follow through with the same steps. The end result of all these numbers may prove fruitful.

We will determine the miles mass for Saturn as we did with the other planets. This effort brought to light a problem with the data I was using for the moon Janus. The two sources of data provided erroneous results that had me stumped. I found after having sent the manuscript off to the publisher that a source advised that JPL on the Web had some accurate data for the moons so I went there to take a look and found better more accurate data than I had been using which fit into the flow of things much better. As always I am dependent upon the data others have provided and when it is off there is no way to make it work with my formulas although the possibility to rework interpolated numbers may solve the discrepancy.

Janus data from original two sources.

Mean orbit Radius miles	Sidereal days	OVMPS	Saturn Mass in miles mass
1- 105,327.3	.749	10.226	11,015,228.04
2- 94,093	.6092	11.23	11,866,341.

The result for miles mass for Saturn was so far out of line with

the other satellites I continued to think about it while the book was with the publisher. The corrected data is used in the chart as reported below with the erroneous figures being discarded.

SATURN

Saturn's Satellites	Semi major axis miles	Sidereal days	OVMPS	Saturn's mass miles
Janus	94,137.55	.695	9.85	9,133,840.0
Mimas	105,580.40	.942	8.922	9,202,059.38
Enceladus	147,893.20	1.370	7.85	9,114,590.20
Tethys	183,313.00	1.888	7.06	9,136,979.84
Dione	234,267.80	2.737	6.22	9,063,446.35
Rhea	327,477.80	4.518	5.27	9,094,175.00
Titan	758,108.00	15.95	3.456	9,054,793.43
Hyperion	919,672.00	21.28	3.14	9,084,267.13
Iapetus	2,212,184.00	79.33	2.027	9,089,264.55
Phoebe	8,078,199.98	550.5R	1.067	9,196,941.81

R=retrograde revolution, i.e. clockwise

The total is 91,170,357.69. The average is 9,117,035.8. As always I must accept the published data as accurate and in this instance I had to rework everything to overcome the in accurate data for Janus, The published data recites that Saturn's mass is 95.17 times that of the Earth. Using the above figures of 9,117,035.8 / 97,022 = 93.96, times as massive in relative mass miles. The general pattern of the results of the various moons seems to follow the results for Jupiter. All in all the results for all of the satellites are reasonably close. My conclusion is that Saturn is only about 94 times as massive as Earth and that the published mass is always greater when Newton's Laws are used. So we still want to learn why. As I did with Jupiter I decided to do a second chart using the data provided in "The Atlas of the Solar System" except for the relative mass data. For our Sn I will use 9,117,035.8, For Janus I will use the data from the JPL Web site. I would use more of that data except I would have to rewrite the entire book to do so.

Saturn's satellites

	Mean orbit Radius	Period days	OVMPS actual	OVMPS using Sn	Saturn's mass
Janus	94,137.55	.695	9.85	9.84	9,133,840
Mimas	115,332	.9424	8.90	8.99	9,135,448
Enceladus	147,955	1.3702	7.85	7.85	9,117,35
Tethys	183,126	1.8878	7.05	7.05	9,101,820
Dione	234,578	2.7369	6.23	6.23	9,104,652
Rhea	327,602	4.5175	5.27	5.27	9,098,457
Titan	759,102	15.9454	3.46	3,46	9,087,665
Hyperion	921,154	21.2776	3.15	3.15	9,140,506
Iapetus	2,212,184	79.3308	2.03	2.03	9,116,189
Phoebe	8,047,130	550.337	1.06	1.06	9,041,755

The hypothetical Sn number for Saturn's miles mass works really well. The first results for Janus, in my opinion, is highest because the orbital data for Janus was wrong but not so wrong as to make it apparent until some method was used to compare the vitals with the balance of the satellites of Saturn.

There are some indicative results for Jupiter's satellites as well and all of this will be discussed later when I discuss my theory concerning the potential for Relative Mass of the planets and satellites.

URANUS

Uranus is an unusually situated planet because the equator of Uranus is inclined about 98 degrees to the planet's orbit. Furthermore the planet has a retrograde, clockwise, rotation. This bears some comparison with the planet Venus. Tip any normal planet 180 degrees or so and it will have a retrograde rotation.

When I originally got involved with Uranus I was thinking in terms of the five satellites that are the largest and best known because they were discovered, except for Miranda, over 100 years ago. Now along comes Voyager 2 and we have a lot of little fellows to add in the mix if I feel like it, which I didn't, at first.

I was hoping to find some pattern with Uranus that would help to explain why Uranus appeared to be measured as less weight massive than Neptune. There are some curious circumstances when the figures for the satellites in clockwise (retrograde) rotation are compared to the figures for the satellites in counter clockwise ("normal") orbits.

Uranus Satellites	Mean orbit radius miles	Sidereal days	OVMPS	Uranus miles mass
Miranda	80,782.00	1.414	4.15	1,394,373.91
Ariel	118,687.40	2.520[R]	3.43	1,392,345.54
Umbriel	161,564.00	4.144[R]	2.84	1,298,758.84
Titania	270,930.40	8.706[R]	2.26	1,387,618.62
Oberon	362,276.20	13.46[R]	1.96	1,387,917.21

R=retrograde revolution, i.e. clockwise.

These are the kind of results I expected to see for each of the planets when I first came up with the miles mass idea. This should tell us something. Even though the last four satellites are in retrograde revolution the results for comparative mass are very similar for all. It is also important to note that Uranus itself is in retrograde rotation. This is the unique object in our Solar System because not only is it in retrograde rotation but it also has satellites. As mentioned previously, Venus rotates retrograde and has no satellites.

All 5 divided by 5, = 1,372,202.7 / 97,022 = 14.14 times the Earth miles mass. The published data states that Uranus is 14.58 times the mass of the Earth. This is a slight difference, which must be accounted for by the different methods used here to measure mass. It is my view that Uranus is potentially more massive than Neptune. Uranus is more voluminous than Neptune. Uranus is in an orbital position that one would expect to be larger than Neptune because it is like stepping-stones from the largest Jupiter to the next largest Saturn to Uranus and then to Neptune. This area of discussion will require a separate chapter. For completeness I am providing a chart for the 10 small moons. It is a constant source of fascination for me and I want to see how the numbers work out. The data was taken from "The Atlas of the Solar System". Theses are tiny objects in small orbits. For a comparison of mean orbital velocities I will show published data and use 1,372,203 as Un.

Uranus	Mean Orbit	Orbit days	OVMPS actual	Miles mass Uranus	OVMPS by Un#
Cordelia	30,741.28	.330	6.774	1,410,627.45	6.681
Ophelia	33,428.83	.372	6.535	1,427,619.13	6.407
Bianca	36,770.10	.433	6.175	1,402,066.89	6.108
Cressida	38,388.23	.463	6.029	1,395,367.67	5.979
Desdemona	38,946.87	.475	5.962	1,384,383.79	5.935
Juliet	39,988.33	.493	5.898	1,391,050.20	5.858
Portia	41,065.22	.513	5.821	1,391,455.60	5.780
Rosalind	43,461.34	.558	5.664	1,394,278.73	5.619
Belinda	46,765.32	.622	5.468	1,398,237.42	5.417
Puck	53,440.40	.762	5.100	1,389,984.80	5.067

All of the smaller moons produce a miles mass number

for Uranus higher than was projected by the larger moons in the previous chart. Even so the mean orbital velocities compare favorably between the actual calculated from the published data and the orbital velocities calculated the old way using the Un number arrived at from the larger moons. The very slightly smaller orbital velocities calculated using the Un number appears to be consistent calling for a very slightly larger Un number. There was no data shown for eccentricity or inclination about the various moons. Actually I think it is really impressive that the Astronomers were able to provide the information they did

The normally revolving satellites produce an overall larger mass miles for Uranus than the retrograde satellites do. If we were comparing the results of each set of satellites with identical planets the size of Uranus we could easily conclude that the planet with the normally revolving satellites was more massive than the planet with the retrograde satellites. This is why I believe that there is a factor of apparent relative mass in the Solar System due to the means of calculating mass. The planets that adhere to the basic choreography of the System will out perform those that do not. I am convinced that such items as retrograde rotation of the planet or the satellites, the degree of tilt off of the perpendicular, and the degree of deviance from the elliptic will all play a part in some degree to produce a greater or lesser measured mass. I am not sure just how to develop this further.

The results follow no apparent pattern. The orbits all conform to what we would expect as to larger orbits producing slower velocities. There is a potential help here, which I plan on testing, shown by the miles mass as to each moon. It shows a potential wave pattern that has been suggested previously in the charts. Nothing in particular is demonstrated, except as mentioned above, by each moon to account for the differences in the mass miles result we see. Let's try Neptune.

Note Voyager disclosed six small objects in orbit around Neptune inside the orbits of the larger moons. I am not reciting the data for those very small moons.

NEPTUNE

Neptune Satellites	Mean orbit radius	Orbit day's	Orbit vmps	Neptune MM
Triton	219,354.20	5.877	2.714	1,616,072.22
Nereid	3,454,984.00	359.881	.698	1,683,282.02

The average miles mass for Neptune is 1,649,677.12 / 97022 = 17.00. The published ratio is 17.24 or 17.26 times that of Earth depending on the source.

The data for Pluto and its moon, Charon, is not clear yet so I doubt I would get any meaningful results, but I will use the most recent data I have found and check it out.

Pluto's mean orbital radius is 3,666,260,000 miles and is very eccentric. Dividing this orbit radius into our Sn number provides for an orbital velocity of 2.94 miles per second. The Moon Charon has a mean orbit around Pluto of about 10,564 miles. On this basis 2.94^2 times 10,564 equal a miles mass figure for Pluto of 91,718. Earth has a miles mass of 97,022. Miles and this would make Pluto about 95% as massive as the Earth and that is so unlikely as to make the effort meaningless. The published data states that the period of revolution for Pluto is 248 years! That seemed wrong until I recalculated the orbit factors and came up with 247.44 years myself.

To bring this recent data together I will list the miles mass for the planets in turn. The various source data that I have been using shows that Mercury's mass is .0558 of the mass of Earth. So using Earths 97,022 miles mass times .0558 I will speculate that the miles mass for Mercury is about 5,413.82. For Venus at .8150 we get 79,072.93.

Planet	Miles Mass	MM/ Earth	Weight mass	Mass Kg
Mercury	5,413.82	.0558	.0558	3.3×10^{23}
Venus	79,072.93	.8150	.8150	4.87×10^{24}
Earth	97,022.00	1.	1.	5.97×10^{24}
Mars	10,319.00	.1063	.1074	6.42×10^{23}
Jupiter	30,378,021.00	313.5	317.89	1.899×10^{27}
Saturn	9,117,035.8	93.88	95.15	5.686×10^{26}
Uranus	1,372,202.70	14.14	14.58	8.66×10^{25}
Neptune	1,663,774.04	17.14	17.26	1.03×10^{26}

It seems possible that a different result could be obtained every time these numerous numbers are recalculated. That is not the way it is supposed to work, but when there are a lot of trailing things after the decimal, it makes long number accuracy some what more elusive. I have tried to be very accurate. Some of the calculations were repeated a number of times. You can do them again if you wish.

There is no reason to expect that the miles mass results are going to equate to the weight mass results obtained by the formula worked out by Newton.

Chapter Seven
Planet Position Numbers

As I said at times previously my original pursuit was for the key to the rotation of the planets. I wanted to be the first to make the discovery of the mechanics behind what the academics see in the heavens. I thought this was a reasonable goal and something that an average guy like me could accomplish. It has been a very long road getting to this point and I am so close to it I am at times afraid I will solve it at any moment – and then what would I do to amuse myself?

By coming up with a measure of the Suns gravitational value expressed in mass miles I was able to see that the Solar System was so precise that it was almost like there were fixed tracks on which each planet and all of its close friends traveled. This point dawned on me as a potential means of finding the cause of the planets rotation. Recall that the planets are not in circular orbits but the effect of their behavior gives us the opportunity to treat them as if they were in circular orbits.

One of my earliest and most obvious observations was that the larger a planet was the faster it rotated. LARGER here is significant and not the same as massive. The proof of that is that Neptune rotates slower than Uranus even though the academics calculate that Neptune is more massive than Uranus. Uranus is larger than Neptune. I had to find a way to take advantage of this uniform effect on the speed of the planets rotation. I already discovered that moving the planets from one orbit to another would change their period of revolution but I did not look to see if it was synonymous with a disclosed change in rotation. I thought that this effect could be broken down into small differences in some meaningful way and came up with a means to test it. If we assume for experiment that the Solar System's great circle could be broken down into multiple tracks then each planet covers an area of many solar tracks. I wanted to see where this would take me. I will fix the orbital velocity of each planet and calculate the factor of its velocity from its nearest point to the Sun and its farthest point to find a number. I will assume that the orbital radius reported for the planets is the planet center.

This book was originally written on 8 ½ by 11 paper with font 14 in Times New Roman. When the publisher tried to format the book to fit then 6" x 9" actual book size many of the original charts would not work as originally laid out. I have had to rewrite and change the layout of data in the charts so that the publisher could make them fit. This means that at times when I think the data is obvious I may eliminate spaces and such to condense the data discussed.

I will use the Suns mass miles divided by the planet's orbit radius. First I will deduct the planets radius in miles from the published orbit radius. I will get the orbital velocity for that position. Then I will add the planet radius in miles and determine the orbital velocity at that position. The slower result will be deducted from the faster result to obtain the difference in the orbital velocities. I will then designate this result as the planets significant number. I will be doing

James J. Wood, Sr.

this later for many of the various planets satellites. This significant number can be very useful for deriving other statistics.

Mercury, equatorial radius is 1,515.6 miles.

Sn/(35,979,060-1,515.6)=884.743934435unsquared= 9.744645475
Sn(35,979,060+1,515.6)=884.669398779unsquared=<u>9.743392523</u>
                                              # .001252952

((1,515.6*6.28320)/#)=7,600,305.4/86,400=87.96, as published.
Or Sidereal period 87.96.
(.001252952/((3.1416 / (87.96 * 86,400))) = 3,031.5.

Venus, equatorial radius is 3,760.7 miles.

Sn(67,235,480–3,760.7)=473.45084307333= 21.758925595564
Sn/(67,253,480+3,760.7)=473.39788273151= <u>21.757708581822</u>
                                          # .001217013742

((3,760.7*6.2832)/#)=19,415,746.4 / 86,400 = 224.72 days.
Published period are 224.6 days.
(.001217013742/((3.1416/(224.72*86,400))) = 7,521.7 diameter.

Originally, when I first tried this approach on Earth with an impressive sequence of interplay of number data I thought that I might get the same results with all of the planets but it did not work out that way. Much of the data does work however and I found it to provide a major help in eventually working out a numerical rotation resolution.

Earth, equatorial radius is 3,963.4 miles

Sn/(92,916,448 –3,963.4)=342.42443543760 = 18.5047138707
Sn/(92,916,448+3,963.4)=342.39523823432 = 18.5039249413
.0007889286

((3,963.4*6.2832) / #)= 31,562,528.36 / 86,400 = 365.30.
Published days are 365.256.

Also note ((3963.4*2)*3.1416)/. 000789 = 31,562,528.36/ 86400=
365.30 days. Dividing the planet circumference as above gives
orbital seconds in a year for all. A reverse search by 86400*
.000789 =68.1696. (7926*3.1416)/68.1696 = 365.26. A further
check for accuracy 92,961,440/ (31562528/6.2832) = 18.5, Earths
OVMPS. Earths equatorial circumference / years seconds
(24,900/31,562,528)=. 000789. Earths surface velocity of rotation
/ years days (.2882 / 365.25) = .000789. I cannot resist these
digressions to show the simple mile mathematics of the System.

Mars, equatorial radius is 2,111 miles

Sn/ (141,617,060–2,111)=224.77086217681 = 14.9923601269
Sn/ (141,617,060+2,111)=224.76416122966 = 14.9921366465
.0002234803

((2,111*6.2832)/#)=59,351,250./86,400=686.93.Published
686.95.

Jupiter, equatorial radius is 44,366.7 miles.

Sn/(483,822,040–44,366.7)=65.79657545319 = 8.1115088271
Sn/(483,822,240+44,366.7)=65.78450940817   = 8.1107650322
.0007437948

((44,366.7*6.2832)/#)=374,787,306.2/86,400= 4,337.8.Pub. 4,337.

**All of these "days" are Earth equivalent days and not
the actual individual planet's rotations.**

Saturn, equatorial radius is 37,284 miles.

Sn/(886,737,800– 37,284)=35.89815682914 = 5.99150705825
Sn/(886,737,800+ 37,284)=35.89513819025 = <u>5.99125514314</u>
.00025191510

((37,284*6.2832)/#)= 929,927,697/86,400 = 10,763. Pub. 10,760.

Uranus, equatorial radius is 17,337 miles.

Sn/(1,784,039,400–17,337)=17.84222002856 = 4.22400521171
Sn/(1,784,039,400+17,337)=17.84187325644 = <u>4.22396416372</u>
.00004104798

((17,337*6.2832)/#)=2,653,768,550.8/86,400=30,714.Pub. 30,700.

Note there is a difference by publishers. Any error in orbit or radius changes the period. I mention this before.

Neptune, equatorial radius is 15,100 miles.

Sn/(2,794,497,940–15,100)=11.39062789301 = 3.37500042859
Sn/(2,794,497,940+15,100)=11.39050479573 = <u>3.37498219191</u>
.00001823667

((15,100*6.2832)/#)=5,202,502,430.5/86,400=60,214.Pub. 60,200.

Pluto, estimated radius is 1,554 miles, 1 source.
This planet data is insecure but I will try this data for the heck of it.

Sn/(3,674,648,900–1,554)= 8.66230448440 = 2.943179315706
Sn/(3,674,648,900+1,554)= 8.66229715787 = <u>2.943178071043</u>
.000001244626

((1554*6.2832)/#)=7,845,039,289/86,400=90,799. Pub. is 90,700.
This came out closer than I would have expected.
Another source says 90,465.

I originally went through this exercise because I felt it was logical that the "tracks" of the planets orbit between the closest to the Sun and the farthest from the Sun would show some idea of the forces on the planet. The leading edge of the planet is in a spot that would normally revolve faster than the spot where the farthest edge of the planet is located. The individual significant number for the planet is the difference of the potential force between the fastest revolution and the slowest. Each planet is unique in this regard as they are located farther and farther from the sun and they are all of different sizes. If we had a planet identical to the Earth between Jupiter and Mars the significant number would be different from the Earth's because the entire duplicate planet is in a slower speed zone altogether. To demonstrate my point I will, for the heck of it, run the same formula as above putting Earth in Jupiter's orbit so we can see the results clearly. Earth's radius is 3,963.3 compared to Jupiter's at 44,366.7

$Sn/(483,826,040-3,963.3)=65.79053688 = 8.111136596$
$Sn/(483,826,040+3,963.3)=65.78945903 = \underline{8.111070153}$
$$\# \quad .000066442$$
$1/.000066442 = 15,054.$

$(483,826,040*6.2832) / (3,963.3*6.2832) = 122,076.56.$
$122,076.56 * .000066442 = 8.11$ miles per second.
$((3,963.3 * 6.2832) /\#) = 374,793,052.6.$
$374,793,052.6 / 86,400 = 4,337.9$ days. Also for Jupiter.

The orbital velocity is properly altered for Earth's orbit radius but otherwise fully consistent with the mean orbital radius of Jupiter. A simple demonstration that the planets size has nothing to do with its orbital velocity, which means the orbital period, is also unchanged. If we compare this position number for the one obtained for the Earth in the correct orbit we find quickly that none of the previous numerical results are forthcoming. The suggestion is that the days, seconds and other criteria are specific to the planets size and location.

I use Earth's significant number, .0007889286, when used to look for meaningful disclosures as .000789 for my convenience. This is also used as a positive number by transposing it, 1/.000789= 1267.4271229404 reduced to 1267.4 for my convenience. My goal in this chapter was to emphasize the very precise nature of the Sun's effect on the planets that provides the opportunity to check every mile from the Sun up to the end of the Sun's mass miles into a specific orbital velocity. An example of this is to take the Sun's radius of 432,000 miles and seek the potential for an orbital velocity at that distance from the Sun. Sn / 432000 = 73,682.671 unsquared provides a potential orbital velocity of 271.4 miles per second. By this method in this Solar System if we know the planets mean orbital velocity we can readily calculate the mean orbital radius of the planets position relative to the Sun by squaring the orbital velocity and dividing it into the Sn number.

To show the curious result numbers will provide in later chapters I note that the Sun's diameter is 864,000 miles for a circumference of 2,714,342.4. This circumference divided by our test orbit velocity of 271.4 miles a second gives us 10,001 as the number of times it goes into the

Sun's circumference. Some numbers, for reasons I have not understood have meaning for this system when divided by 10,000 or 1,000. Another example is the Earth's mean orbital radius of 92,916,448 divided by 271.4 = 342525. That divided by 1,000 gives us 342.525 unsquared is 18.5 that are the velocity of Earth in its orbit around the Sun.

A by-product of this effort, as briefly hit on above, is that the significant number for each planet does provide the planets seconds in orbit based on a full revolution. When we divide that result by 86400 seconds for each Earth day we get the equivalent Earth days for that planets orbit. This is not the actual rotation for the other planets because they do not have 86400 seconds in each of their rotations. In addition, because the mechanics of the Sun-planets relationships being so accurate this method provides the opportunity to double check the academics observations of a planet to confirm if the numbers fit the formula.

These "significant numbers" can be used to manipulate other data. An example is to divide Earths number of .000789 by Earths diameter of 7926.6 and getting .00000000995. Then take 3.1416 divided by .0000000095 that will be 31,564,352.45 miles. That item divided by 86400 seconds provides 365.32. Not a bad estimate of Earth days in a year. As of today I feel I have not discovered all the possible use for these significant numbers. I will do the same thing with the major satellites just to be thorough. At one time I worked out what I considered to be the actual rotational speed for all planets. My preliminary effort based primarily on published data is: [references to published data mean I worked it to miles]

Planet	Orbit radius	Sidereal day's	Orbit vmps	Estimated rotations
Mercury	**35,977,324**	**87.96**	**29.75**	**1.5**
	Published day rotation 58.65 days.			
Venus	**67,232,236**	**224.68**	**21.76**	**0.924**
	Published day rotation 243.01 days.			
Earth	**92,956,955**	**365.25**	**18.5**	**365.4**
	Published day rotation 86,400 seconds.			
Mars	**141,610,227**	**686.95**	**14.99**	**669.62**
	Published day rotation 88,642.44 seconds.			
Jupiter	**483,612,285**	**4,337.**	**8.11**	**10,575.85**
	Published day rotation 35,427.60 seconds.			
Saturn	**886,695,015**	**10,760.**	**5.99**	**25,247.76**
	Published day rotation 36,838.80 seconds.			
Uranus	**1,783,953,320**	**30,700.**	**4.22**	**47,601.18**
	Published day rotation 55,800 seconds.			
Neptune	**2,794,363,106**	**60,200.**	**3.38**	**91,324.52**
	Published day rotation 56,880 seconds.			
Pluto	**3,666,222,305**	**90,700.**	**2.94**	**14,197.58**
	Published day rotation 6.3874 days.			

Chapter Eight
Satellite position Numbers

We have run through some satellite charts for the planets. I have wondered why these various objects form where they do around a planet. I have speculated that there are gravitational waves that determine the stuff that will make a planet or a satellite in a position where we find it today. On second thought it might be possible that the stuff of planets varies a lot causing some to gather together and some stuff to be isolated. There is a possibility that all of the above things are happening during Solar System creation.

It is possible that some Astronomer worked all of this out previously and I am unaware of it and how he or she did it. I saw a shareware program that created Solar Systems that you could manipulate somewhat using a built in formula employing, I believe, Sir Isaac Newton's formulas. The systems created were entirely random as far as I could see. When I used the orbit data for the planets satellites we found that we could get a measure of mass expressed in miles for not appear to play any consistent part in the final result of mass miles for the planet. At first I thought that the planets mass miles showed to be less when the satellite was a more massive one, or a denser one. This was not a consistent showing either.

Knowing that I am anxious to get what is finished published I am concerned that I will not have time to work out a formula to explain the reasons of what I find in the various behavior of the objects in this Solar System. We shall see. For Jupiter the four major satellites produce curious comparisons. Take another look.

Moon	Jupiter's Mass miles miles	mass in kilograms	Mean orbit r a d i u s
Io	30,382,240	8.94×10^{22}	261,982
Europa	30,383,607	$4.8\ \times 10^{22}$	416,897
Ganymede	30,357,280	1.48×10^{23}	664,898
Callisto	30,410,331	1.076×10^{23}	1,170,096

Ganymede is the most massive of the four but shows miles mass for Jupiter less than any of the four moons listed. If we were getting a number for the combined mass of Jupiter and Ganymede we would expect the result to be larger than the others.

Callisto is slightly less massive than Ganymede and produces slightly higher mass miles for Jupiter, which starts to make a point of progression until we compare Io and Europa. Io is shown as twice as massive as Europa yet Europa produces a higher mass miles than Io, as opposed to lesser. I will average the above mass miles results to get 30,383,364.5 as the Jn mass miles number. There is a sort of consistency when you consider that the more weight massive the satellite is the less miles massive the planet will appear to be. The differences are so slight that comparison seems useless. Europa providing 30,383,607 divided by Ganymede at 30,357,280 = 1.00087 times a larger result while Ganymede is 3.08 times as weight massive as Europa.

Amalthea   194   miles radius

112,660-194=112,466 Jn/112,466=270.108477227=16.434977250
112,660+194=112,854Jn/112,854= 269.17982526= <u>16.406700620</u>
.028276629
((194*6.2832/#) = 43,107.7 / 86400 = .498, same as published.

Io   1,132 miles radius

261,982-1,132 =260,850 Jn/260,850=116.457811002=10.7915620
2
261,982+1,132=263,114Jn/263,114=115.455734016=<u>10.74503299</u>
.04652903
((1132*6.2832/#)= 173,041269 / 86400 = 1.769 same as published.
Also (.0465290351/2264) = .00002055169 or X.
 3.1416 / X = 152,863.54/86400= 1.769 days.

Europa   970 miles radius

416,897- 970= 415,927Jn/415,927=73.03690311= 8.5461630636
416,897+970= 417,867Jn/417,867=72.69782011 = <u>8.5263016666</u>
.0198613970
((970*6.2832)/#)=306,861.8/86,400 = 3.551. Same as published.

Europa   1,100 hypothetical radius for comparison.

416,897-1,100 =415,797Jn/415,797=73.07259191= 8.54825081018
416,897+1,100=417,997Jn/417,997=72.68799656= <u>8.52572557407</u>
.02252523611
((1,100*6.2832)/#)= 306,834.5 / 86,400 = 3.551 days.

This was a shocker. This is not quite the surprise it seemed at
first. Recall we are using the same orbital data and the velocities
are narrowed down to the central point of the object. The method
does work to demonstrate the presence of erroneous data in the

three main items that are discussed. I did expect that the use of a larger satellite radius would effect the outcome of the days result.

Ganymede 1,635 mile radius

664,898-1,635 = 663,263Jn/663,263=45.80086632= 6.7676337314
664,898+1,635 = 666,533Jn/666,533=45.76168021= <u>6.7510123701</u>
.0166213613

((1,635*6.2832)/#)= 618,062. / 86,400 = 7.1535 days.
 The published days are 7.155.

Callisto   1,497 miles radius

1,170,096-1,497=1,168.599Jn/1,168,599=25.9952473=5.09855345
1,170,096+1,497=1,171,593n/1,171,593=25.9288165=<u>5.09203462</u>
.00651883

((1,497*6.2832)/#)= 1,442,889.1 / 86,400 = 16.7 days.
The published days are 16.689.

Leda   using 6.2 miles hypothetical radius.

6,897,540- 6.2=6,897,534 Jn/6.897, 534=4.40418561=2.098615165
6,897,540+6.2 =6,897,540Jn/6,897,540=4.40418178= <u>2.098614252</u>
.000000913

((6.2*6.2832)/#)= 42,678,703. / 86,400 = 494 days.
The published days are 238.7. This is an example of a glaring error due to an inappropriate choice of the radius of the moon at the least but it could very well be faulty orbital data.

((2.99*6.2832)/.0000009123)= 20,582,149. / 86,400 = 238.2 days.
Very close to actual days published. The #, .000000913 remains the same for each alternative moons radius.

Leda   again using a radius of 3 miles.

6,897,540-3=6,897,537Jn/6,897,537=4.404183696= 2.098614708
6,897,540+3=6,897,543Jn/6,897,543=4.4041798652= <u>2.098613796</u>
.000000912

((3*6.2832) / #) = 20,650,985.46 / 86,400 = 239 days.
Published days are 238.7 & 239. This is the best data so far.
Previous comments regarding conflicts in the published data still prevail. In some instances I just eliminate the faulty data.

Himalia   53 miles radius

7,127,458-53=7,127,405Jn/7,127,405=4.262143094= 2.064495845
7,127,458+53=7,127,511Jn/7,172,511=4.262079707= <u>2.064480493</u>
.000015352

Published days for one revolution are 250.6.
This provides 251.06. Close enough.

Lysithea   7.5 miles radius.

7,276,594-7.5=7,276,586 Jn/7,276,586=4.17476243 =2.04322354
7,276,594+7.5=7,276,601Jn/7,276,601=4.17475383=<u>2.04322143</u>
.00000211

Published days for one revolution are 259.2.
This provides 258.988 days. Close enough.

Elara   25 miles radius.

7,297,100- 25=7,297,075 Jn/7,297,075= 4.16304067= 2.040353074
7,297,100+25=7,297,125 Jn/7,297,125=4.16301214= <u>2.040346084</u>
.000006990

Published days for one revolution are 259.7.
This provides 260 days. Close enough.
The probable radius for Elara is 24.96 miles to get 259.7 days.

The next three satellites are shown from two different sources in the chart above because of different details in the published data. The orbit details vary but the days recited are not so far apart. We know that any change in the orbit radius will make a difference in the orbital velocity of the satellite. The difference presented in the data should not have a major effect.

Ananke   6.2 mile radius

12,862,980-6.2=12,862,973.8Jn/12,862,973.8=2.3616638= 1.53677058
12,862,980+6.2=12,862,986.2Jn/12,862,986.2=2.3616615 =<u>1.53676984</u>
.00000074

This works to 608.69 days and the published days are 631.
This data source has questionable data.

Ananke   second published data.

13,173,680-6.2=13,173,673.8 Jn/13,173,673.8=2.30596418=1.51854015
13,173,680+6.2=13,173,686.2Jn/13,173,686.2=2.30596201=1.<u>51853943</u>
.00000072

The published days from source are 631.1. I get 630.88,
Close enough. So this data looks OK.

Carme   9.3 miles radius

13,888,290-9.3=13,888,280.7Jn/13,888,280.7=2.18731322= 1.47895680
13,888,290+9.3=13,888,299.3Jn/13,888,299.3=2.18731029= <u>1.47895581</u>
.00000099

This works to 683 days. The published days are 692.
 This data looks faulty.

Carme   second published data.

14,043,640-9.3=14,043,630.7Jn/14,043,630.7=2.16311726= 1.47075397
14,043,640+9.3=14,043,649.3Jn/14,043,649.3=2.16311439= 1.47075300
                                                    # .0 0000097

This works to 694.4 days. The published days are 692.5. This data is more accurate. It seems any error is most likely the orbit radius.

Pasiphae   11.2 miles radius

14,478,620-11.2=14,478,608.8Jn/14,478,608.8=2.0981312= 1.44849273
14,478.620+11.2=14,478,631.2Jn/14,478,631.2=2.0981279 =1.44849167
                                                    # .00000106

This works to 726.9 days. The published days are 744.

Pasiphae   second published data.

14,602,900-11.2=14,602,888.8Jn/14,602,888.8=2.08027476= 1.44231576
14,602,900+11.2=14,602,911.2Jn/14,602,911.2=2.08027156=1.44231465
                                                    # .00000111

This works to 736.2 days. 738.9 published days. This is better.

Sinope   8.7 miles radius.

14,727,180-8.7=14,727,171.3Jn/14,727,171.3=2.06271933 = 1.43621702
14,727,180+8.7=14,727,188.7Jn/14,727,188.7=2.06271689= 1.43621617
                                                    # .00000085

This works to 745.7 days. The published days are 758. A moon radius of 8.84 might work. We cannot be certain if an error is in the orbit radius or the moons radius.

I dislike typing so many numbers as was done here but on the other hand I did not want to leave any doubt about how the results were obtained. We have done enough of these to show the method works for checking the three items in the formula. If the orbit data and the moon radius are correct the days will always fit the observed days in the orbits revolution around the planet. In reverse, if you know the orbit radius and the days, you can determine the probable size of the satellite. Otherwise you can determine the days by orbit velocity.

The last satellite, Carme, shows us that the there is a balance of sorts built into the method. The closer orbit needed a larger satellite to make it work and the larger orbit wanted a smaller satellite to provide an accurate result. When we compare the two published sources I think we can conclude that the second source most closely represents the situation.

Another interesting thing about the numbers is the progression getting smaller and smaller as we go farther from the planet. The exception is Amalthea and Io with Io showing a larger number than Amalthea following which all the others falls in line. These satellites are somewhat close together but have significant size differences that may play a part. To provide an example of the days, distance and moon radius relationships, I could use the spot for Himalia and instead of the proper radius of 53 miles substitute an object of 310 miles radius for the formula. We narrow the data down to the objects center so we can get the same days with a larger object.

Himalia   53 miles radius

7,127,458-53= 7,127,405, Jn/7,127,405 = 4.26214309 = 2.06449584
7,127,458+53= 7,127,511 Jn/7,172,511 = 4.26207970 = <u>2.06448049</u>

.00001535

Published days for one year are 250.6 and this provides 251.06. Close enough.

Test item   310 miles radius

7,127,458- 310 = 7,127,148 Jn/7,127,148= 4.26229678 = 2.064533066
7,127,458+ 310 = 7,127,768 Jn/7,127,768=4.261926033 - 2.064443274
.000089792

(310*6.2832)/.000089792=21,692,266.57/86,400   =251.06   days   as above.
We got the same result, as above meaning the orbit data is incorrect.

This method shows that (.000089792/.00001535) = 5.84903. 5.84903 * 53 = 310, so we see that the radius of the moon is not a major factor in determining the resultant days when all of the other ingredients remain the same. However, if the radius is incorrectly recited in the formula we saw earlier that the days result may be erroneous. I have manipulated the moons radius while leaving the orbit radius alone. I am sure that very slight variations in the orbit radius will work the same way that the moons radius does.

We can note that when the orbit distance remains the same a larger moon increases the resultant days and a smaller moon decreases the days. If that is correct then a larger orbit will increase the days and a smaller orbit decrease the days. This may seem like child's play to the academics. If so, it does not detract from the fact that the formula provides a means of double-checking observed data for accuracy. To be consistent I will run this through for the outer planets as well.

We have some extra space here and I want to emphasize my concern over some of the charts that have been really sized down to where small type is used and spaces may be eliminated for the sequence. Many years ago I tried being a salesman, especially while going to law school at night. As is my way I obtained a large number of books from the public library on how to be a good speaker and a good salesman. Two things stuck with me over the years. One is that you do not interrupt your self to apologize for errors while you are speaking. The other was some smart fellow that taught you to sell the sizzle and not the steak. The smell will sell it best; the prospect of it will sell it best. These were sound advice and I wish I could recall who the authors were. Unfortunately, a slip of the written thought once written is not easily overlooked and I am sure that my readers will note the complications of which they will behold and wonder if I am still selling the sizzle and not the product. You will see it as it was meant to work.

Chapter Nine
Saturn's Satellite Numbers

Now that we have made a preliminary exploration of Jupiter's major moons we should check to see what will develop for Saturn. The same steps for each of the moons will be gone through. The end result of all these numbers should prove fruitful. I am doing this to assure myself that my simple formula for the orbit, days and moon radius works for all of the moons which this time are farther from the Sun revolving around Saturn.

The average miles mass for Saturn including all but Janus is 9,118,897.5. For my convenience I will use the value for Saturn at 9,118,897 miles mass as the Sn number. The radius data for each of Saturn's moons was obtained from an Internet source that I did not discover until later, http://enclyclopedia.thefreedictionary.com/. As I did with Jupiter I decided to do a chart using the data I have used before. For our Sn I will use 9,118,897.

SATURN

Saturn's Satellites	Semi major axis miles radius	Sidereal period days	Vel	Saturn mass in miles
Janus	94,137.55	.695	9.85	9,133.840.00
Mimas	115,580.40	.942	8.92	9,202,059.38

Enceladus	147,893.20	1.370	7.85	9,114,602.20
Tethys	183,313.00	1.888	7.06	9,139,239.44
Dione	234,267.80	2.737	6.22	9,076,584.10
Rhea	327,477.80	4.518	5.27	9,098,868.17
Titan	758,108.00	15.95	3.46	9,057,453.55
Hyperion	919,672.00	21.28	3.14	9,084,266.76
Iapetus	2,212,184.00	79.33	2.03	9,097,521.66
Phoebe	8,078,199.98	550.5^R	1.07	9,199,482.27

R=retrograde revolution, i.e. clockwise

This chart goes through the individual characteristics for each of Saturn's moons. We will be using the moons radius deducted and added to the orbit radius and the results divided into the miles mass number for Saturn to get OVMPS squared then unsquared. Because some of the moons are not spherical I must interpolate a radius that fits the other factors of each orbit as best I can. When I can find a published radius for a moon it is always in kilometers. I simply convert it into miles by multiplying by .62137. Saturn Sn = 9,120,391.75

Janus   **Sized at 97 x 95 x 77 km.**
          A hypothetical radius of 59 miles is used.

94,137.55-59= 94,078.5 Sn / 94,196.5 = 96.9444485 = 9.84603906
94,137.55+59=94,196.5Sn / 94,196.5 = 96.8230428 = <u>9.83987006</u>
00616900   ((59*6.2832)/#)= 60,092.2 / 86,400 = .695 days,
The published days are .695. I originally used two textbook

sources for the Janus data. Both were way off. A real problem attempting to correct the data until I went to JPL for the data.

Mimas

radius reported as 121.2 miles and as 209 x 196 x 191 km.

115,580.4-121.2 =115,495.2Sn/115,495.2=78.9547747= 8.8854993
115,580.4+121.2=115,701.6Sn/115,701.6=78.8139273= <u>8.8777208</u>
.00777885
((121.2*6.2832)/#)= 97,894.9 / 86,400 = 1.13 days.
The published Earth days are .942. The radius is probably closer to 101 miles to fit the orbit and the formula. Reverse ((86400*.942)*#)= 633.08 / 6.2832 = a probable radius of 100.76 miles.

Enceladus radius reported as 155.3 miles and as 256 x 247 x 245 km.

147,955-155.3=147,799.7 Sn/147,799.7=61.6976725 = 7.85478660
147,955+155.3=148,110.3Sn/148,110.3=61.5682872= <u>7.84654620</u>
.00824040
((155.3*6.2832)/#)= 118,418.54 / 86,400 = 1.37 days,
The same as published at 1.37 days.

Tethys

radius reported as 326.2 miles and as 536 x 528 x 526 km.

183,126-326.2=182,799.8 Sn/182,799.8= 49.8846142 = 7.06290409
183,126+326.2=183,452.2 Sn/183,452.2=49.7072125= <u>7.05033421</u>
.01256978
((326.2*6.2832)/#)= 163,048.6 / 86,400 = 1.887 days.
The published days are 1.888.

Dione

radius reported as 348 miles or 560 km.

234,578 -348 = 234,230 Sn/234,230 = 38.931381 = 6.2395017067
234,578+348 = 234,926 Sn/234,926 = 38.816042= <u>6.2302521676</u>
#.00924953903

((348*6.2832)/#)= 236,395.95 / 86400 = 2.736 days.
 The published days are 2.737.

Rhea   radius reported as 475.3 miles or 764 km.

327,602-475.3=327,126.7 Sn/327,126.7=27.87573591= 5.27974771
327,602+475.3=328,077.3Sn/328,077.3=27.79496630= 5.2720931
                                                                    .00765455

((475.3*6.2832)/#)= 390,147.6 / 86400 = 4.516 days.
The published days are 4.518.

Titan   radius reported as 1,590 miles
or 1,600 miles.

759,102-1,590 = 57,512 Sn/757,512= 12.03795781=3.4695760280
759,102+1,590 =760,692Sn/760,692=21.98763428=3.4623163179
                                                    # .0072597101

((1590*6.2832)/#)= 1,376,127.7 / 86,400 = 15.927 days.
The published days are 15.945.

Hyperion   radius reported as 90 miles
and 185 x 140 x 113 km.

921,154-90 = 921,06   Sn/921,064=9.9003950865 =3.14648932725
921,154+90= 921,244 Sn/921,244= 9.8984606684=3.14618191915
                                         #   .00030740810

((90*6.2832)/#)= 1,839,535.13 / 86,400 = 21.29 days.
The published days are 21.277.

Iapetus   radius reported as 447.4 miles
and 446.14 miles

2,212,184-447.4=2,211,737 Jn/2,217.366=4.12295743= 2.0305066
2,212,184+447.4=2,212,632Sn/2,212,632=4.12128971= 2.0300959
                                                    # .0004107

((447.4*6.2832)/#)= 6,844,546.3 / 86,400 = 79.22 days.
The published days are 79.331.

Phoebe   **radius reported as 70 miles and 115 x 110 x 105 km.**

8,047,130-70=8,047,060Sn/8,047,060=1.133196161= 1.064516867
8,047,130+70=8,047,200Sn/8,047,200=1.133176446=<u>1.064507607</u>
                    # .000059260

((70*6.2832)/#)= 7,428,498.3 / 86,400 = 86 days.
The published days are 505.5r. (JPL says 548.21 days)
Reverse ((86,400*505.5)*#)=2,585.9. 2,585.9/6.2832= 411.56 mile radius, which is impossible. Phoebe is in retrograde revolution and quite possibly traveling slower than expected. This example is not strong but never the less indicative of my contention that retrograde orbits show less mass.

The hypothetical Sn number for Saturn's miles mass works really well. Jupiter's satellites raise questions as well and all of this will be discussed later when I discuss my theory concerning the potential for Relative Mass of the planets and satellites.

Now that I have gone through the long-winded method of obtaining the "significant" number for the planets and the satellites thus far I will note for comparison that the same result can be obtained by dividing the planets circumference by the seconds in a years orbit. Earth then is ((86400*365.25) / (7926.6*3.1416)) = 1,267.3. 1/1,267.3 = .000789. A lot easier but not discovered until after I had learned what the significant numbers were. While I am on the subject I should share another interesting item. The significant number divided by the planets diameter gives another precise comparison for consideration. Earth with, .000789 divided by 7,926.6 provides .00000009953. However 3.1416 divided by the seconds in Earths year will give the same result. (3.1416/31,557,600) = .00000009955. This interesting play of numbers should prove meaning full in my searching. I will continue in the next chapter with the laborious way of finding my significant number to check the accuracy. The numbers show clearly how numerically mechanical the Solar System really is.

You may recall that Earth's year days of 365.25 times Earths significant number like, (365.25*.000789), equaled .28818. .2882 miles are the equatorial surface velocity of Earth's rotation as it revolves around the Sun. I have not as yet explored the other planets or satellites in this manner to see if some useful rotation data will be disclosed.

This topic is fully discussed in a later chapter.

Chapter Ten
Uranus, Neptune & Pluto

Uranus is an unusually situated planet because the equator of Uranus is inclined about 98 degrees to the planet's orbit. Furthermore the planet has a retrograde, clockwise, rotation. This bears some comparison with the planet Venus.

When I originally got involved with Uranus, I was thinking in terms of the five satellites that are the largest and best known because they were discovered, except for Miranda, over 100 years ago. Now along comes Voyager 2 and we have a lot of little fellows to add in the mix, if I feel like it, which I didn't, at first. There is more repetition here but I feel compelled to continue with the search for the "significant numbers" as was done in the past chapters for Jupiter and Saturn. I will run it out for all of the known satellites of Uranus. The R means retrograde revolution.

Uranus Satellites	Mean orbit radius miles	Period days	OVMPS	Uranus MM
Miranda	80,782	1.414	4.15	1,394,373.91
Ariel	118,687.40	2.520^R	3.43	1,392,345.54
Umbriel	161,564.00	4.144^R	2.84	1,298,758.84
Titania	270,930.40	8.706^R	2.26	1,387,618.62
Oberon	362,276.20	13.46^R	1.96	1,387,917.21

These are the kind of results I expected to see for each of the planets when I first came up with the miles mass idea. This should tell us something. Even though the last four satellites are in retrograde revolution the results for comparative mass are very similar for all. It is also important to note that Uranus itself is in retrograde rotation. This is the most unique object in our Solar System because not only is it in retrograde rotation but it also has satellites. As mentioned previously, Venus rotates retrograde and has no satellites.

All 5 1divided by 5, = 1,372,202.7 / 97,022 = 14.14 times the Earth miles mass. The published data states that Uranus is 14.58 times the mass of the Earth. This is a slight difference, which must be accounted for by the different methods used here to measure mass. It is my view that Uranus is potentially more massive than Neptune. Uranus is more voluminous than Neptune. Uranus is in an orbital position that one would expect to be larger than Neptune because it is like stepping-stones from the largest Jupiter to the next largest Saturn to Uranus and then to Neptune. This area of discussion will require a separate chapter. For completeness I am providing a chart for the 10 small moons. It is a constant source of fascination for me and I want to see how the numbers work out. Theses are tiny objects in small orbits. For a comparison of mean orbital velocities I will show published data and use 1,372,203, based on the 5 larger satellites as Un. This will cause the larger miles mass results to show a slower mile per second orbital velocity and the lesser mass miles results to show a faster miles per second orbital velocity.

This planet has the most extreme showing for differences in the miles mass calculations. The moons in retrograde orbits show generally less miles mass for Uranus. This falls in line with my thinking that retrograde revolution or retrograde rotation and planet tilt is a factor in the apparent mass of a Solar System body. I think that measuring the mass of a planet by the orbital periods of its satellites, whether for weight or for my mass miles ratios, will cause the erratic satellites to show smaller values for the parent planet than those more upright satellites that are in normal, more uniform, orbits and orbit in the manner the Sun rotates. This circumstance, I think, will carry over to reflect erroneous data for Planets orbiting the Sun as well. This is a preliminary thought based on the limited data Uranus provides. The confusion of the mass miles result is apparent when you compare the data. The five major moons average 1,372,203 mass miles for Uranus. The total of 15 satellites provide an average of 1,388,455.5 miles mass for Uranus and the 10 moons in more normal orbits provide an average of about 1,396,581.75 miles mass for Uranus. We can reasonably speculate that if Uranus rotated counter clockwise with all of the satellites in counterclockwise orbits Uranus would be somewhat larger giving Uranus a larger mass image no matter what method was used.

Uranus Satellites	Mean orbit radius miles	Orbit days	Miles ovps	Mile mass Uranus	Miles by Un
Cordelia	30,741.28	330	6.774	1,410,627.45	6.681
Ophelia	33,428.83	.372	6.535	1,427,619.13	6.407

Bianca	36,770.10	.433	6.175 1,402,066.89	6.108
Cressida	38,388.23	.463	6.029 1,395,367.67	5.979
Desdemona	38,946.87	.475	5.962 1,384,383.79	5.935
Juliet	39,988.33	.493	5.898 1,391,050.20	5.858
Portia	41,065.22	.513	5.821 1,372,203.00	5.780
Rosalind	43,461.34	.558	5.664 1,394,278.73	5.619
Belinda	46,765.32	.622	5.468 1,398,237.42	5.417
Puck	53,440.40	.762	5.100 1,389,984.80	5.067
Miranda	80,782.00	1.414	4.15 1,394,373.91	4.121
Ariel	118,687.40	2.520[R]	3.43 1,392,345.54	3.400
Umbriel	161,564.00	4.144[R]	2.84 1,298,758.84	2.914
Titania	270,930.40	8.706[R]	2.26 1,387,618.62	2.250
Oberon	362,276.20	13.46[R]	1.96 1,387,917.21	1.94

R=retrograde revolution, i.e. clockwise.

As mentioned above all of the smaller moons that are in counterclockwise revolution produce a miles mass number for Uranus higher than was projected by the larger moons that are in clockwise rotation. Even so the mean orbital velocities compare favorably between the actual calculated from the published data and the orbital velocities calculated my way using the Un number arrived at from the larger moons. There was no data shown for eccentricity or inclination about the various moons. Actually I think it is really impressive that the Astronomers were able to provide the information they did. I included the moon radii because that data was available and I would be using it in the next sequence. I was also curious to see if potential differences in the size of the moons showed something useful and that endeavor remains to be tackled.

The orbits all conform to what we would expect as to larger orbits producing slower velocities. There is a potential help here, which I plan on testing, shown by the miles mass as to each moon. It shows a potential wave pattern that has been suggested previously in the charts. Nothing in particular was demonstrated by each moon to account for the differences in the mass miles results we see. Miles radius was used from a source I failed to note at the time. The 2nd note is from the radius data provided from the Web site I mentioned earlier. I made a guess for them for Miranda that they show as 240 x 234 x 233 km. and for Ariel that they show as 581 x 578 x 578 km, In an effort to make a worthwhile comparison of the satellites as was done before I will use a Miles Mass number for Uranus based on the 10 small moons. Un = 1,396,581.75

Uranus Satellites

Cordelia   Radius of 16.16 miles or 12.43 miles.

30,741.28-16.16=30,725.12Un/30,725.12=45.4540698= 6.74196335
30,741.28+16.16=30,757.44Un/30,757.44=45.4063065=6.73842018
                                          # .00354317
((16.16*6.2832)/#)= 28,657 / 86.400 = .3316 days.
Published days are .330. .0016 over.
Reverse ((86,400*.330)*#)= 100.14 / 6.2832 = 15.94 miles radius.

Cordelia using moon radius of 12.43, from 2nd note,

30,741.28-12.43=30,728.85Un/30,728.85=45.448552= 6.74155415
30,741.28+12.43=30,753.71Un/30,753.71=45.411813 = <u>6.73882880</u>
                        #   .00272535
((12.43*6.2832)/#)= 28,656.99 / 86,400 = .3316 days like before.
Size of moon radius is not the issue here.

This example suggests that the Un is in error and that could be because I am using a Un that was not derived from the small satellites which are now the objects being tested. This is unique to Uranus due to the difference in the orbits of the large moons and the smaller moons. It may be that the recited orbit data may be very slightly off the actual. However when I take the full orbit of 193,153.6 divided by my OVMPS of 6.681 I get the days at .3346! The calculated orbit velocity of 6.774 will provide the .330 days recited. My view is that I may have a used an invalid mass miles number.

Ophelia radius of 18.64 miles or 13.04 miles.

33,428.83-18.64=33,410.19Un/33,410.19=41.8010717=6.46537483
33,428.83+18.64=33,447.47Un/33,447.47=41.754480= <u>6.46177073</u>
                        #   .00360410
((18.64*6.2832)/#)= 32,495.96/86,400=.376 days. The published days are .372. This is close enough that I will not run out the alternate radius from the other source.

Bianca radius of 26.10 miles or 16.77 miles.

36,770.1-26.10=36,748 Un/36,748 = 37.34086208= 6.110716986
36,770.1+26.10=36,796.2Un/36,796=37.29194862= <u>6.106713406</u>

.004003580

((26.1*6.2832)/#)=40,961.2 / 86,400= .474 days. .433 days published. Taking the orbit circumference divided by 6.175 MPS and 86,400 will provide .433 days.

	Orbit Radius	days	ovms	Uranus MM	Un ovms
Cressida	38,388.23	.463	6.029	1,395,367.	5.979
Desdemona	38,946.87	.475	5.962	1,384,383.	5.935
Juliet	39,988.33	.493	5.898	1,391,050.	5.858
Portia	41,065.22	.513	5.821	1,372,203.	5.780
Rosalind	43,461.34	.558	5.664	1,394,278.	5.619
Belinda	46,765.32	.622	5.468	1,398,237.	5.417
Puck	53,440.40	.762	5.100	1,389,984.	5.067
Miranda	80,782.00	1.414	4.15	1,394,373.	4.121
Ariel	118,687.40	2.520^R	3.43	1,392,345.	3.400
Umbriel	161,564.00	4.144^R	2.84	1,298,758.	2.914
Titania	270,930.40	8.706^R	2.26	1,387,618.62	2.250
Oberon	362,276.20	13.46^R	1.96	1,387,917.21	1.94

R=retrograde revolution, i.e. clockwise.

I decided not to repeat any more of the testing for the balance of Uranus satellites. If any one is interested in testing the accuracy of the published data they have plenty of examples on how to do it. Uranus has a lot to teach us due to its unusual satellite patterns. If I was to project the days, orbits and moon radius again I would use a separate Un number for the moons in retrograde revolution and a separate Un number for the moons in normal revolutions.

Let's try Neptune.

Note Voyager disclosed six small objects in orbit around Neptune inside the orbits of the larger moons. I am not reciting the data for those very small moons.

Nn = 1,649,677.12

Neptune Moon	Orbit radius	Orbit days	Orbit vmps	Orbit Neptune miles mass	Moon radius
Triton	219,354.20	5.877	2.714	1,616,072.	845.1

219,354.2-845.1=218,509.1Nn/218,509.1=7.54969527= 2.74767088
219,354.2+845.1=220,199.3Nn/220,199.3=7.49174552=2.73710531
01056557
((845.1*6.2832)/#)= 502,569.7 / 86.400 = 5.86 days.
 The published days are 5.877.

| Nereid | 3,454,984.00 | 359.88 | 698 | 1,683,282 | 52.5 |

3,454,984-52.5=3,454,931.5Nn/3,454,931.5=.47748475= 69100271
3,454,984+52.5=3,455,036.5Nn/3,455,036.5=.47747024= 69099221
.00001050
((52.5*6.2832)/#= 31,416,000. / 86,400 = 363.6 days.
The published days are 359.881.

The average miles mass for Neptune is 1,649,677.12 / 97022 = 17.00. The published ratio is 17.24 or 17.26 times that of Earth depending on the source.

Pluto

The data for Pluto and its moon, Charon, is not certain yet but I think that the Atlas of the Solar System has some more recent data and I will use it. I may not get any meaningful results, but I will use the most recent data I have found and check it out.

Mean orbital radius is about 3,666,083,000 miles and is a very eccentric orbit.
Dividing this orbit radius into our Sn number provides for an orbital velocity of 2.95 miles per second. The Moon Charon has a mean orbit radius around Pluto of about 10,564 miles. We are advised that Charon has an almost circular orbit and that Charon makes a 360-degree orbit around Pluto in 6.2 Earth days. Which is recited as the rotation period for Pluto? 10,564 * 6.2832 = 66,375.72 miles for a full orbit of Charon around Pluto. 66,375.72 / 86,400 tells us that Charon orbit velocity is .768 mps. That is the velocity I used to get Pluto's miles mass figure. On this basis $.768^2$ times 10,564 provides a miles mass figure for Pluto of 62,309. Earth has a miles mass of 97,022. This would make Pluto about .6422 as massive as the Earth. The Web data shows Earth with a mass of 5.9736×10^{24} Kg. Pluto is shown as 1.25×10^{22} Kg. The Solar System Atlas provides Pluto with a mass of 6.6×10^{23} Kg. and that seems more real to me. By comparison this shows the relative Pluto/Earth mass as .1104. A more favorable showing for

Pluto's mass calculation than above and less mass than my efforts demonstrate. Typically, the miles mass data is less than the quoted weight mass in kilogram comparisons between the Earth and the other planets.

Charon   radius of 373 miles (best estimate)

10,564 -373=10,191 8,113.152/10,191= .797610950=   .89224968
10,546+373=10,919 8.113.152/10,919=.743030680=   86199227
                                                    # .03025741
((373*6.2832)/#)= 71,947.64 / 86,400 = .8327 days.
This is the formulas suggestion.

I have no published days except that the Atlas states Pluto rotates in 6.2 days. We can do it in reverse with .8327 days times 86400 giving us 71,945.28 times # provides 2,176.88 miles circumference, divided by 6.2832 provides a radius of 346 miles for Charon. That is as good an estimate as the others offered. The published data states that the period of revolution for Pluto is 248 years! That seemed wrong until I recalculated the orbit factors and came up with 247.44 years myself.

Mercury, as noted in chapter three has a mean orbit radius of 35,977,324 miles. The estimated rotations in that full orbit circumference are only 1.5. That is almost standing still in terms of rotations. The day's based on Earths days of 86,400 seconds in a day could be about 89.96 rotations but for unknown reasons Mercury does not come close. Mars has about 88,642.44 seconds in each rotation. Mars is larger than Mercury so we should expect to see a somewhat faster rotation, which means less seconds in a Mars day than in a Mercury day. There is no final formula yet to calculate the proper rotations that Mercury should have. That is also true for Venus and all of the other planets. You should recall that for Earth the days in a year times by the E# of .000789 equaled .2882, and that **is** Earths equatorial surface velocity of rotation in miles per second.

Earths' diameter of 7926.6 * 3.1416 = 24,902.2 / 86,400 = .2882. The best method I have so far says that Mercury's Earth days of 89.96 * .001252952 = .1127 SV. The circumference of 9,522.8 / .1127, the estimated surface velocity, equals 84,497 potential seconds in a Mercury day, or one rotation. This is less than Earth seconds. I must check this further. With an orbit circumference of 226,052,722 miles traversed at 29.75 miles per second it takes about 7,598,410 seconds per full orbit and that provides for possibly 90 days in one Mercury year. So why only 1.5. On this basis Venus would have about 86,395.6 seconds in one Venus day. This method is nowhere near accurate but there is a potential for Venus to rotate some 220 times in one full orbit and we see 1 retrograde rotation in a full year, how can this be? Venus was flipped over a long time ago and is just about now getting it' rotations down to zero so it can reverse rotation and join the pack. This will start the cooling down process as normal rotations slowly catch up where they belong. This may not have been the best place for this discussion but there it is. You may get more of this viewpoint later on in the book. To bring this recent data together I will list the miles mass for the planets in turn. The various source data that I have been using shows that Mercury's mass is .0558 of the mass of Earth. So using Earths 97,022 miles mass times .0558 I will speculate that the miles mass for Mercury is about 5,413.82. For Venus at .8150 we get 79,072.93.

Planet Mass	Miles Mass	<u>MM</u> Earth	<u>Pub Mass</u> Earth	Weight Kg.
Mercury	5,413.82	0558	0558	3.3×10^{23}
Venus	79,072.93	.8150	8150	4.87×10^{24}
Earth	97,022.00	1	1.	5.97×10^{24}
Mars	10,319.00	.1063	.1074	6.42×10^{23}
Jupiter	30,378,021	313.5	317.89	1.9×10^{27}
Saturn	9,133,840.	93.88	95.15	5.686×10^{26}
Uranus	1,372,202.	14.14	14.58	8.66×10^{25}
Neptune	1,663,774.	17.14	17.26	1.03×10^{26}

Chapter Eleven
Developing Details

I have worked the planets statistics many times in many different ways in the expectation of discovering some surprises. There are many surprises as I see it but not final answers in every case depending on what you may be seeking. As I have explained previously I have been looking for the answer to planet rotations in the form of a mathematical resource.

You will recall that I worked out what I considered to be a significant number for each of the planets. I also discovered a rather simple way of getting the same number for each of the planets by dividing the planets circumference by the planets seconds in a year. This was a real surprise for me and caused me to speculate further on the relationship of a planets circumference to the planets orbit. This discovery offered more surprises when I learned that planets actual days in a year multiplied by the significant number would disclose the planets surface velocity of rotation. This chart will disclose these results. Keep in mind that only the earth has 86,400 seconds in a day and that number is only useful when you are looking for Earth comparisons. I went through the calculations for each planet position number in chapter seven. The Earth manipulations showed some comparisons of interest and possibly they would work for the other planets.

I am aware that I entered the planet radius in two parts, plus and minus, when arriving at the planets number. The mean orbital radius was also entered. However, neither the actual days of the orbit or the seconds in the orbit were disclosed to the formula.

Mercury and Venus give us very little data to play with. Mercury is estimated to have only 1.5 rotations in its entire trip around the sun. We can conclude that fact is not normal and that Mercury should have rotations equivalent to its size, as do all the other planets. Venus is worse due to the retrograde rotation completely out of sync with the main theme.

Recall the larger a planet is the faster it rotates. Larger being by volume or surface area and not by mass.

I will list the data from 3 different sources for comparison of the data.

	Mercury	Venus R	Earth	Mars
1, Rotation period	88 days	30 d(?)	23h56m	24h37m
2, Rotation period	58.65days	243day	23h56m	24.6hr
3, Rotation period	59days	244.3da	23h56m	24h37m22s

	Jupiter	Saturn	Uranus	Neptune
1,	9h53m	10h26m	10h42m	15h48m
2,	9h.8	10h.2	16?	16?
3,	9h50-55m	10h14-38m	12h	15h48m

I am not sure which of the above observations are the earliest. They are all in doubt for the large planets due to the difficulty of observation with heavy cloud cover and long distance.

From the very beginning, some 30 years ago, I was looking for the academics rendition of the Solar System secrets. Many of the numbers changed as time progressed.

Mercury Orbit radius	Significant number	Full orbit seconds	Earth days	Times rotate
35,979,060	.001252952	7,598,777.	87.96	1.5

$((3,031*3.1416)/7,598,777.47) = .001253203.$ $(87.96*.001252952) =$ .1102 svmps of planet equator for a day of 89,264.7 seconds. 7,598,777.47 / 89,264.7 seconds in a day would give us 85.13 real days; possibly.

Venus				
67,235,480	.001217013	19,414,244.	224.7	0.924

$((7,521.4*3.1416)/19,414,244.85)=0012171078$
$(224.7*.0012170137) = .2734$ svmps of equator for a day of 86,427 seconds. 19,414,244.85 / 86,427 seconds would give us 224.6 real days; possibly.

Earth				
92,961,440	.000789	31,572,719.	365.256	365.256

$((7,926.6*3.1416)/31,572,719.98)= .000789$
$(365.256*.000789) = .2882$ svmps of equator.
31,572,719.98 / 86,400 seconds would give us 365.4 days.

I leave a lot of numbers trailing which can put the result off a small fraction. You can approach the issue any way you wish and you will find that the System is precise. The planet Mercury is small, less than half the Earths diameter and shows that Earth at 86,400 seconds divided by the Mercury potential of 27,507 seconds equals 3.141. Earth at 86,400 divided by Venus at 86,407 seconds in a day is turning slower than Earth and the suggested surface velocity also shows Venus turning slower than Earth. No ratios are evident yet.

Mars

141,617,060   .000223480   59,357,263.   686.95   669.7

Observed rotation of 24.62 hours.
((24.62*60)*60)= 88,632 seconds in a day.

59,357,263.39 years seconds divided by 88,632 equals 669.7 days.
((4,222*3.1416) / 59,357,263.39) = .000234576

669.7*.0002234576)= .14965 svmps of planet equator.
59,357,263.39 / 88,721 = 669.7 or 669.7 days specific for Mars.

Jupiter

483,635,620   .0007437948   374,677,276   4,336.5   10,620.1

Observed rotation of 9.8 hours.
((9.8*60*60) = 35,280 seconds in Jupiter's day.

374,677,276 years seconds divided by 35,280 = 10,620.1 days.
((88,730*3.1416) / 374,677,276) = .00074398471
(10,620.1*.00074398471)=7.901 svmps at for a day of 35,280 secs.
374,677,276 / 35,280 = 10,620.1 rotations.

Saturn

886,737,800   .0002519151   930,097,181. 10,765 25,329.4

Observed rotation of 10.2 hours. (10.2*60*60) = 36,720 seconds in Saturn's day. 930,097,181.6 year of seconds divided by 36,720 equals 25,329.4 rotations.

((74,564*3.1416)/930,097,181.60 = .00025185568
(25,329.4*.00025185568)=6.379 esvmps for   a day of 36.720 seconds.
930,097,181.6 / 36,720 = 25,329.4 rotations or 25,329.4 days specific for Saturn.

Uranus

1,784,039,400 .0000410479   2,656,274,018 .30,74461,487.8
Observed rotation of 12 hours (12*60*60) = 43,200 seconds.
2,656,274,018 / 43,200 = 61,487.8 rotations.

Neptune

2,652,794,498 .0000182366   5,194,539,132 60,122   91,324
Observed rotation of 15.8 hours. (15.8*60*60) 56,880 seconds.
5,194,539,132 / 56,880 = 91,324.5

This is an example of some of the comparisons I ran out to test the rotation possibilities. In many cases I would write a basic computer program so I could manipulate the data over and over again by simply changing the measurements that I thought would work. Needless to say none of them worked although some were close to the published data. The published data provides no assurance that it is accurate and that makes the problems solution more illusive.

I think it is necessary to reason the factors before jumping onto some calculation. The Solar System is a very busy place. The entire System is revolving with the Galaxy as it turns on its center. The sun is rotating at about 1.25664 miles per second at the equator. Sunlight leaves the sun in a straight line that continues to the planets in an ever expanding and speed-increasing pattern until it reaches each planet. Even though the orbital velocity of each planet is different I have not decided that the planets velocity plays a part in the planets speed of rotation but it is a point to be considered. Another item of note is that the sun being a sphere is expanding, as a sphere meaning the expansion is forward and sideways from all points on the sphere. If we were to see the Sunlight stopping at Earth we would see a horrendous sphere that goes above and below us. Dividing our radius of 92,961,440 by the suns of 432,000 gives us 215.2. We must think in terms of the surface of the sun expanding and not the radius. I ran out the surface of a sphere with a radius of 92,916,440 miles and divided it by the surface of the sun and got a surface 11,576.524 times the surface of the sun this sideways expansion suggests to me that Sunlight plays a part in rotation.

So, why do the planets rotate at all? Our Moon is said to make one rotation as it travels around the Earth exposing the same face to us all the time. I think the planets rotate, and originally started rotating in a counter clockwise direction for a combination of reasons. We know that the size of the planet is a factor because large planets rotate faster than small planets. I will not concern myself now with the satellites rotations, whatever they are, since I have no data to work with.

If we assume a circle with a radius of the orbit up to the point at which the Earth revolves around the sun we can envision a hypothetical line going completely around the sun with the sun at the center. If we then place the Earth, or any planet, with its closest edge to the sun on that line we can visualize the planet rolling along that line. I call this the rolling ball effect that is due to the pull of gravity holding the planet in this hypothetical position, while other factors cause the actual rolling. This would be like the planet having an axle running up and down through from one pole to the other. The planet would rotate as if a long shaft was connecting the axle to the center of the sun. If there were no other contributions to the final result we could expect the planet to simply revolve around the sun with out continuous rotations. We must next add the effect of sunlight constantly expanding and exerting pressure on the planets. We must then consider the density of the sunlight in the locale where the planet is located and also the fact that the sun is turning potentially causing the light to have a moving effect similar to the beam of a navigational lighthouse beam of light. The rotation of the suns equator could pull the planet along while at the same time exerting a pull on that side of the planet closest to the sun. This in combination would account for the rolling ball effect but not necessarily for the speed of rotation that I doubt would involve the size of the planet as we find it to be. The final contributing cause is the effect of sunlight on the planet. This results from the side of the planet facing the sun to receive more pressure on the right front side than on the left back side. There is no specific vacuum formed because the entire planet is in a vacuum but the effect can be compared to the effect of an airplane wing causing the airplane to get the lift to fly. In the planets situation it is simply the difference in Sunlight pressure, and rotating gravity from the sun on the planet, causing the planet to rotate along the line of the rolling ball point of revolution. As I suggested previously all planets in this Solar System should rotate counter clockwise and if they do not then something major occur to change them to what we see now?

I want to repeat that the System's effects on the planets by the actions of Sunlight and gravity cause the rotation of the planets and when the planet fails to respond in the normal manner these forces can cause compression and excessive heat on the planet. Many years age, while having a coffee break with some other lawyers and an Administrative Law Judge the subject of the Voyager space craft heading towards Uranus came up. Some knew that I was interested in such things and asked me what I thought about it. My primary comment was that I was sure they would discover that Uranus was giving off more heat than it was getting from the sun. I forgot about the conversation entirely. Later when NASA published their preliminary findings, and before I learned of it, The Judge, who had an excellent memory, mentioned to me that I was correct in my assumption about Uranus. That is a very self-serving recitation but it is true none-the-less. The Judges initials are LD. My prediction was based on the known fact that Uranus was in retrograde rotation.

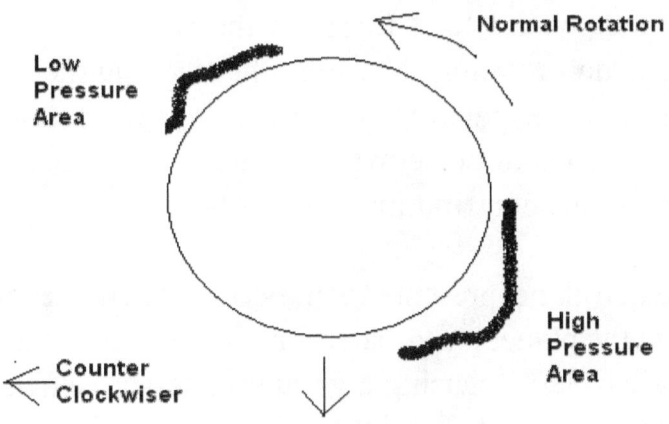

The sun is over in this direction.

The size of the planet plays a part in the speed of rotation because the surface area of the planet is in contact with the light and gravitation waves meaning more surface means more pressure applied and that energy is converted into greater surface velocity. Recall my comments about the fast lane closest to the sun and the slow lane farthest from the sun based upon the extremes of the planets surfaces. That fact could cause confusion with my contention here wherein one might think the difference in the speeds of the tracks would cause the planet to rotate clockwise with the fast track taking charge over the slow track. Not so. The tracks do not appear to contribute in any known way to the planets rotation or the planets speed of rotation although the planets orbital velocity may very well play a part.

To reiterate, we have a planet in orbit around the sun being held in place by the suns gravitational pull. This relationship causes the planet to assume a continuous circular (elliptic) orbit around the sun. The suns pull on the closest side of the planet would tend to cause the planet to remain motionless, non-rotating. The sunlight reaching the planet at great speed is expanding continuously as it reaches and passes the particular planet while the suns gravitation field rotates the space contributing to the rotations.

Because each planet presents an individual face to the forces exerted on the planet it is greater for a large planet than it is for a smaller planet causing a greater or lesser velocity of rotation for the planet. To convert these controlling factors into a comprehensive formula for predicting the probable rotational velocity for a planet of a given size is the next step, which has not been deciphered by me, or any one else, yet. I am satisfied that I have outlined the ingredients that must be measured in the formula.

Astronomers do not consider that sunlight has a mass of its own. However we know that sunlight has a pressure and, as I have stated elsewhere, it has a density that is specific to its distance from the sun. We know that the tails of comets always point away from the sun. The prospect of sunlight's location specific density does not appear to be a factor in the rotation of the planets otherwise we could expect to find a planets orbital radius and position playing a part in the planets rotation. Size is the one uniform ingredient that relates to all planets.

When I ran out the math for the significant numbers for the planets I found that the Earth days in a year divided by the reverse of the significant number, or multiplied by the significant number, gave me the surface velocity for the planet in miles per second velocity at the planets equator. I have used that method with success to determine the surface velocity for every planet in this System as follows:

Days Seconds

Mercury
(87.96 / (1 / .001252952)) =.1102 svmps.
(3031.2*3.1416) / .1102) = 89,264.7 seconds per rotation.

Venus
(224.7 /(1 / .0012170137)) = .2734 svmps.
((7521.4*3.1416) / .2734) = 86,427.3 seconds per rotation

Earth
(365.25 / (1 / .000789)) = .2882 svmps.
((7926.6*3.1416) / .2882) = 86,406. Slightly off 86,400.

Mars
(669.7 / (1 / .0002234803) = .1496 svmps.
((4222 * 3.1416) / .1496) = 88,662. Seconds per rotation.

Jupiter
(10,620 / (1 / 0007437948) = 7.92 svmps.
((88,730*3.1416) / 7.92) = 35,285.3. Seconds per rotation.

Saturn
(25,329 / (1 / .0002519151) = 6.38 svmps.
((74,564*3.1416) / 6.38) = 36,716.3. Second per rotation.

Uranus

(61,487 / (1 / .0000410479) = 2.524 svmps.

((34,672*3.1416) / 2.524) = 43,155.9, Seconds per rotation.

Neptune

(91,324 / (1 / .0000182366) = 1.665 svmps.

((30,200*3.1416) / 1.665) = 56,982.7. Seconds per rotation.

In theory we should be able to derive a formula that would permit us to predict the rotational velocity of any spherical object in orbit around the sun. This would most likely require a constant that could be applied to a planets size or other planet configuration in the expectation that the result would disclose the equatorial surface velocity either in miles per second or seconds per mile. The seconds in one rotation, one day for the specific planet, are very specific. I have not found a common denominator for it yet.

Chapter Twelve
A Little Digression

After typing all of those tedious numbers I wanted a break so I indulged myself with this.

I should apologize, I suppose, to all the Astronomers and other professionals that have so diligently studied the Solar System for some of the outlandish views I will be offering on the nature and workings of the System. In general new ideas should be welcome and if greeted with objective review may be shown to contain some merit for further study. I am not concerned with criticism that may be leveled at me even though I take the details herein seriously.

I view the Solar System as a sort of living entity that is capable of change affecting its physical characteristics. I believe that planets can appear to be more or less massive based on their relative position in relationship to the Sun. The means we use to measure or weigh the planets contributes to the prospect for error. I will contend that Planets are capable of limited expansion or compression and that this effect, when it occurs, can in turn be related to a planets position relative to the Sun, and mass that may be added to the object.

The Sun at the center is dependent on the action of the planets to maintain its generating power. I have, since childhood, held that opinion and have always wondered why Astronomers doubted that the other Stars in the

Galaxy had planets. I know that my "belief" will not be persuasive. I see the Sun as a generator of power that requires the tugging of the planets to sustain the generating action. If Stars are born without planets we must assume that either all-excess substance is dispersed outwards far enough to escape the Stars gravity or that all substances are some how sucked into the forming Star. Neither of these scenarios is logical. When we assume the stuff is blown away how far would that be? If we assume that the stuff is sucked up into the formed star how do you reconcile that with the theory that the ignition of the Star blows away the loose stuff from the central part of the Solar System to collect again a little farther away? This particular item is not important to my overall views but I felt it was worth a mention because there are planets everywhere and there is going to be life almost every where not on every planet but on many of them. While I'm at it I may as well turn you off some more by contending that our Solar System had life on at least three planet at about the same time life found a home here on Earth.

I learned quite early in this pursuit that the relationship between the planets and their satellites was quite different from what the Sun planet relationship is. The Sun emitting energy can be expected to influence the planets by more than simply the effects of gravity. The planets appear to have only gravity to motivate the satellites. This leaves open the question of whether the Sun plays a part in the motion of the satellites. Science tells us that as the planets formed from rings of matter revolving around the Sun as did the satellites form in like manner around the planets. This does seem to be true but it also is a cause for me to wonder about it. For instance, when a dust ring reached some certain point, possibly relating to a mass relationship to the Sun or a parent planet, the dust collected into a ball. Why would that ball assume revolution around a planet? If we assume that the dust cloud itself was already in motion around the planets dust ring then we have a collision course during the formation of the planet and the planet does not have a true dust ring circling the sun. If we accept a ring forming a planet it would be a ring going all the way around the sun in the eventual orbit of the finally formed planet and the satellite join it later.

Logic compels me to believe there were rings after ring after rings with some being small and some being larger. We know from the final product that these rings were most likely moving at different speeds similar to the speed of orbit that the planets themselves finally displayed. We must also conclude that the rings comprised a complete orbit around the Sun for each object that was a part of the creative process. There is no room for satellites forming around planets during the creative process. That would be a collision course dissecting the larger ring of the eventual planet. The eventual marriage of the satellites to the planets, in my opinion, had to take place after the spheres were created and interacted with the planets gravity. The proximity of the substance field of forming satellites, which were forming nearby, and possibly some wanderers that joined up later.

Where does the motivating force come from? The Solar System we see defines itself as all-inclusive. The separation of the planets and the separation of the satellites are apparently all subject to some defining gravitational wave that causes separation of the dust and other matter into specific rings prohibiting the forming of a larger planet that would absorb the additional matter that forms into satellites. This discussion does not propose to answer the questions posed. It simply looks for issues.

In our Solar System there are curious divisions between the planets. At first the sequence of Mercury, Venus and Earth seem normal going from the smaller planet to the larger planet in sequence. Then we have Mars that is much smaller than Venus and Earth followed by Jupiter, which is the largest of all. What was working the original formation of the Solar System to cause this result? I do not accept a random placement of matter in the system.

It seems that because the planet is revolving around the Sun as it gathers the stuff that planets are made of, the excess dust, shaped as a ring around the proto planet, gets its momentum from the effect of the planet's larger dust ring in revolution around the Sun. This observation may seem childish to you at first blush. We have, I think, a mystery buried in this planet satellite relationship. What factors cause the separation of some of the matter to ring the planet without being absorbed along with the rest of the matter into the planet? At what point is the satellite too far and there after the next and the next. The lack of satellites within the inner area of the terrestrial planets seems quite natural. The present theory that I believe states that all of the excess stuff that did not adhere to the terrestrial planets was blown out of the inner area by the suns ignition, a Small Bang.

I do not believe that Earth's Moon formed with us originally. The Moon was either an interloper that came later or was the first planet from the Sun and was relocated. I feel this speculation is supported by the relative size of the Moon as compared to the Earth and by the obvious appearance of the Moon's surface. The Moon's gray bombarded surface is the result of it being someplace else when it was formed. The look of the surface suggests the prescience of water in the past and that could be the result of the moon originating, or spending time, in environments where meteoroids are formed followed by the escape of the melting ice when the Moon got deeply into our System. The prospect that we had the moon here before Venus was relocated would account for some past water. To me the logical conclusion is that the Moon was an individual member of the inner Solar System or possibly a visitor that caused a great deal of upheaval getting where it is now. This is consistent with the lack of Moons for Mercury and Venus. Mars does not have moons in the true sense because Phobos and Deimos are not spheres but rather large rough chunks of stuff more consistent with something Mars picked up long after the planet was completely formed. Technically we have no moons in the inner, terrestrial, area of the Solar System.

The planet Mars seems an unlikely neighbor for us. Mars equatorial diameter is about 4,222 miles. Mercury is about 3,031 miles in diameter. Venus is slightly smaller than Earth at 7,521 miles. Earth at 7,926 miles diameter is only 404.6 miles larger in diameter than Venus is now. I believe that the Earth went through an expansion over what it was originally and intend to discuss this change in another chapter. Between Mars and Jupiter we have the asteroids. The asteroids represent unique objects that are neither planet nor satellites. They revolve around the Sun as do the planets but they are on their own so to speak. Astronomers tell us that all of the asteroids crushed together would not provide enough matter for a planet like Mars. Some are spherical and some are like large rocks. Their origin is uncertain. They are found roughly between 185,922,880 and 278,884,320 miles from the Sun (from the Atlas of the Solar System). My view is that they represent the remains of a planet that was demolished, very likely as a result of a collision with Earth's Moon at a distant time before our memory. I will suggest a name for this planet as "Atlan" which means nothing except a means by which to refer to it. They list 35 asteroids with estimated diameters and recite that there are about 3,000 of them in smaller sizes. I tried to total the numbers provided and used a hypothetical number for the extra 3,000 to arrive at a potential diameter for Atlan of 23,265 miles. This is larger than the astronomers would allow but I feel that a larger planet is required to balance the system. This is my hypothetical reconstruction of the Solar System which has the Moon, Mercury, followed by Mars and then by Earth to be followed by Venus, then Atlan, Jupiter, Saturn, Uranus and Neptune. I do not want to play with Pluto. This comes about because I think Venus was originally in an entirely different orbit and was a planet larger than the Earth with abundant water and life forms. Venus is now a very hot planet because it is in retrograde rotation generating immense heat due to electromagnetic and gravitational friction. Venus was tilted 180 degrees as a result of the Moon, or other much larger body's interference, into this System. It was originally rotating counter clockwise as do the other planets and simply went backwards by being turned 180 degrees. In the process Venus orbit was changed from being more distant from the Sun to being closer to the Sun that in turn caused a kind **of** compression which itself contributed to the heat generated within the planet. More on this thought later.

As of now we have worked our way to Jupiter. Why would the System's largest planet find itself about 13% of the way towards the extreme of the Solar System? From Jupiter on out the planets all demonstrate a consistently diminished size. My guess is that when the System formed and the rings of matter were still forming, the forces at the center forming the Sun and other forces at the edge pulled the matter in the rings into the shape to make planets as we see them today. I do not envision any major changes in the system from the original formation of the large planets except for the tilting of various planets and the location of planets off of the elliptic line. If my speculation is correct some large object, like a binary star, was in play while all of this was taking shape. If so it could be dark now but still there somewhere near the outer edge. The alternative explanation, at least one of them, is that when the Sun exploded to life the loose matter from the inner area was pushed out to the areas of the large planets contributing to their larger size. This does not hold sway with me because, if true, why would the planets diminish in size instead of growing in size. My money is on the presence of a large massive object being the cause for what we see.

From time to time you will note that the numbers I use are not exactly the same. This is not of any importance and is not done for any purpose. It may be that a formula used very slightly different data or data from a different source in the mathematics or some similar thing. The results will normally be accurate in spite of very small differences.

Elsewhere I explain how I used the data to come up with a Solar System probable minimal diameter that I designate as my Sn. This was done by taking the orbital velocity of Earth in its orbit at 18.5 miles per second squared for 342.25 and multiplying that by Earths mean orbit radius around the Sun. This provides the point in space where an object orbiting the Sun will have an orbital velocity of 1 mile per second. This is 31,816,052,840 miles that I refer to as the Sun's mass expressed in miles mass. This in turn was turned into a circumference that for my own convenience I rounded to a circumference of 200,000,000,000, miles. This might suggest that the massive object is in that vicinity of about 31,830,914,183 miles from the Sun.

If there was an object in an orbit 200 billion miles long it would take 2,314,814, Earth years to make one revolution around the sun traveling at one mile per second. To have had an effect on the formation of the Solar System and offset the sun's gravity to put enough stuff for Jupiter and the other large planets where they are, if any effect there was in real time, planet x must be very massive but not necessarily a failed sun. Things like this are best left to those that use Newton's Laws to calculate their results.

I will offer my reconstructed Solar System.

Planet Name	Mean Orbit miles radius	Orbit year seconds	Orbit mps velocity	Yr/ 86,400 = rotations
Mercury	35,979,060	7,597,777	29.75	87.0
Mars	67,235,480	19,414,244	21.76	224.7
Earth	92,961,448	31,572,722	18.5	365.42
Venus	141,617,060	59,360,127	14.99	687.03
Atlan	186,281,000	89,551,704	13.07	1,036.47
Jupiter	483,822,040	374,839,783	811	4,338.42
Saturn	886,737,800	930,142,060	5.99	10,765.53
Uranus	1,784,039,400	2,656,274,018	4.22	30,743.91
Neptune	2,794,497,940	5,194,789,780	3.38	60124.88

The above chart is not very meaningful for my speculations. The hypothetical placement of the planets is still subject to correction and the rotations are simply the Earth equivalent if the planet had 86400 seconds in each rotation, which obviously they do not. The result says that each planet has an orbit that has x equivalent Earth rotations. The published data provided frequently contain this type of results.

From the above you can see that I believe it was at least the Moon that was the cause of the major changes in our Solar System. There must have been more due to something larger than we have knowledge of. There may have been some contribution from Pluto and Charon but I suspect they were victims more than perpetrators. The evidence is long gone for any finger pointing. The results are still evident today. Uranus is almost on its side causing a retrograde rotation and, as I hope to show, a diminished apparent mass. Earth is tilted, has expanded over its original size and has a large moon. Venus is tilted about 180 degrees causing a retrograde rotation, a relocation of the planets orbit and a compression of the body of the planet and magnetic conflict, causing extreme heating. These are the major most obvious changes.

This view of mine has nothing to do with theories like that of Immanuel Velekovsky as expressed in his "Worlds in Collision". I read his book some time ago after reading how there was a serious attempt by astronomers and other academics to have the book barred. It is claimed that the publisher was pressured not to print the book. That sounded so extreme to me that I had to read it for myself. His book contends that there was or were major disruptions in this Solar System by the planet Venus (I think) that was born like a meteor out of the planet Jupiter. I did not really grasp his basis for this view except he quoted many varied myths and legends. His research on the subject was amazing to me. He was not concerned with the mathematics or the physics of the Solar System as I see it. I mention this in passing because I can see some readers tossing my views in the trash as was done to his. I hope to provide something more.

This chapter was injected here to jump ahead a little with some thoughts that will be considered far out by my readers. Why are they far out? What is there about our planets that would eliminate the prospect of expansion or contraction? When we check the published data we find that just about every planet and every satellite has been assigned a different density. From what science has determined the entire Solar System is made of essentially the same stuff. Given those facts why the large variance in planet densities? Why couldn't planets be manipulated by significant events just as you would move a heavy crate here on earth? Some Sage is alleged to have said, "give me a large enough lever and I will move the Earth", or something of the kind. We are not really certain just what keeps the earth and the other planets in their places so why condemn the prospect of movement to other locations?

.

Chapter Thirteen
The Speed of Light

Every body talks about the speed of light but nobody has ever shown a mathematical formula to explain why light travels at this famous talked about speed. Is it known?

Talking about Light, I am, for good reason, reluctant to open a thought. So many brilliant people have, over the centuries, expressed some very concrete views that I get a little dizzy at the prospect of expressing my views. One major contributor was Sir Isaac Newton for his studies of the prism and for his analysis of the make up of light. Are we talking about light bullets or is it light waves. How about both? Does light have mass? The consensus says no. Why does light travel as it does through out the Universe? Where does all the push come from? I don't think we know all the answers. Dr. Albert Einstein made the contribution that the speed of light is finite. It sets the bar on all things traveling in space. I recall Dr. Carl Sagan, during his Television series, making a major point of the "fact" that the speed of light was fixed. I think he employed some people on bicycles and offered that if the speed of light was not fixed you could not tell the respective positions of the bicycles. Everything would be confused. At the time I was thinking in terms of light years distance lighting affects not my neighbor's bicycle. Recently a very few academics have questioned the fixed speed of light and speculate that travel faster than light may be possible.

Just today, February 3, 2005, as I was writing from my old notes, I took a break to see what might be new on the Internet about light speed. I lucked upon an article by an academic that I think was written about June 21, 2001. This stuff is way over my head but I could glean from his expertise and his review of many papers on the subject that there was, and are some differences of opinion concerning the speed of light in the very distant past and what it is measured at now. The article was fascinating even though I could not grasp the value of the scientific details. Simply said there are some academics that feel light traveled much faster at the start following the Big Bang and they are searching for why it slowed down along the way to what they measure now. This points the way to some possible variation, over long time periods, for a possible change in the speed of light. The article does not favor any view that light now travels at anything other than at the fixed measured speed.

My intuition tells me that the speed of light is a variable-albeit very limited in scope. I am aware that academics have done various measurements and conclude that light travels at approximately 186,281 miles per second, everywhere, and all of the time. I think that the speed of light at 186,281 miles per second seems arbitrary. Why that precise speed? I have never seen an explanation as to why light would top out at that speed as opposed to something more or something less. I have a reason and an explanation for why we measure the speed of light here at Earth and stress that these measurements have been made from here on the Earth and subject to Earths environment.

Does light blast off from the Sun at an instantaneous 186,281 miles per second? If we were to measure the speed of light 100 miles from the Sun is it traveling at 186,281 miles per second? I think not. As strange as the explanation may sound to you, I will share my own view of light.

Light to me is like an explosion that must either gain speed or lose speed as it expands. This expansion is the means by which light gets to travel the Universe, always getting weaker as its substance expands to make that extra mile. If this thought is correct light must be most dense at the source and less dense everywhere else. This simple observation is confirmed by the heat generated by a light source and that the heat diminishes as we get farther from the source. Density, in this regard serves to describe my view of light as an expanding energy source much like the effect of a gas being released from compression. There should be a progression as light gets farther from the source becoming less and less dense, less powerful but capable of greater speed as it travels through space.

I cannot prove this contention by physical measurement. It would be ideal if the speed of light could be measured at the location of the planet Mercury and possibly at the planet Neptune. I am aware that measurements are made under circumstances believed to be beyond Earth. I fault this method because either way the measurements are made from Earth with the light making a round trip from some reflecting light source, like another planet, to us here on Earth. Starlight is the same way except the light travels only one way but with a speed that will be peculiar to the particular source.

Try to travel mentally with me from the Sun towards the Earth. We are surrounded by light. As we leave the Sun's surface we are at lights greatest density. We are enveloped in light that will eventually expand from an original circumference equal to the Sun's circumference of 2,714,342.4 miles to about 200,000,000,000 miles circumference at the minimum probable edge of the Solar System. I think that is about 73,682.67 times less dense; less illuminating and providing less heat at that point in space. Let's suppose we are now entering the Solar System from space. We are at a weak point wherein light is faint and not at all dense. As we approach the Sun we encounter progressively more dense light.

This real difference in density would not be much of a hindrance to our solid body but how would it affect light that was being projected from space, or reflected light from a planet? There is no doubt that light will travel at different speeds through different substances. I believe this is referred to refraction of the light. Glass or water will slow the speed of light. I believe that the light entering the inner parts of the Solar System would be compressed to conform to the particular density at each point within the Solar System that it entered. This means that any light measured at Earths location, no matter what the source, will measure the same as the speed of light as we get it from the sun. A simple analogy would be equal to an effort to insert gas at 10 pounds pressure per square inch into a container containing gas that was at 40 pounds pressure per square inch. It will not work. The issue remains as to what pressure or speed does our Sun blast off its light? Some years ago I hit upon the prospect that the orbit velocity of a planet had some inclusion brought about by the speed and density of light. The result was the measured speed of light which refers to light speed in the vacuum of space. This prompted my working up a chart which I contend shows the speed of light from the Sun and what we measure it at here on the Earth.

The numbers I have worked out may not satisfy the dedicated academics, but I will offer them as the only way available to me to advance my theory. *No other theory exists*.

I start with a Sun of 864,000 miles in diameter divided by 2000 which = 432. I then start with the somewhat arbitrary appearing conclusion that $432^2 = 186,624$. This is offered as the probable speed of light in this Solar System at the point of the minimal radius of the System or about 31,830,914,183 miles from the sun. The orbital velocity of each planet squared when deducted from the 186,624 will provide the speed of light at that planets location. This example will provide the probable speed of light at each planet. It may seem impossible that the speed of light could be anything but fixed and constant, but I contend it is not. Light is flexible and can be compressed or expanded with lenses and mirrors, much like a gas can with pumps. When light is compressed it gets hotter, as with a 'magnifying glass', much like a gas. The Solar System is full of light but is cold due to the expansion of light and then, on contact with an object, gets hotter due to the effects of the collision impact. I find it interesting and strange that it seems completely normal to people that light gets hotter when compressed but do not see the effect of light impacting on surfaces as demonstrating that light has substance. A white metal panel will get warm in the Sun on a hot summer day but a black metal panel gets too hot to handle. Why?

I think it is due to the extent of impact the colors cause. I think this is otherwise explained as the black panel "absorbing" more heat from the sunlight. A choice of wording. White reflects some of the light thereby reducing the degree of impact. Black reflects about none taking the heavy hit.

The next chart is intended to show the speed of light starting at the sun and progressively getting faster as it heads out into the Solar System. When it reaches the Earths current position it is traveling at about 186, 281 miles per second.

Examples:

Planet	Orbit Radius In miles	OV miles	OV^2 miles	186624 less OV^2
Sun at exit	Equator	271.4	73,682.67	112,941.32
Mercury	35,977,324	29.74	884.46	185,739.54
Venus	67,232,236	21.76	473.49	186,150.51
Earth	92,961,440	18.505	342.243	186,281.75

The measured speed of light here at Earth is about 186,281.75 mps.

Planet	Orbit Radius In miles	OV miles	OV^2 miles	186624 less OV^2
Mars	141,610,227	14.99	224.70	186,399.30

Jupiter	483,612,285	8.11	65.77	186,558.23
Saturn	886,695,015	5.99	35.88	186,588.12
Uranus	1,783,953,320	4.22	17.81	186,606.19
Neptune	2,794,363,106	3.37	11.35	186,612.64
Pluto	3,666,222,305	2.94	8.64	186,615.35
Unknown	31,830,914,183	1.	1.	186,623.
Unknown	50,000,000,000	.6366	.4052	186,623.59

Many books on astronomy will round out the measured speed of light at about 186,000 miles per second. I recall a more accurate recitation at 186,281.25 or so. On the Internet I saw a conclusion that the speed of light is 299,792,458 meters a second. Divide that by 1,000 and multiply by .62137 and you get 186,282 miles per second. I do not think it is a coincidence that the measure of the speed of light at Earths location, compared to my prediction, is precisely as the academics have measured it. By the same method the initial blast of light from the Sun's surface should be traveling much slower in miles per second after which it rapidly and progressively gains speed as it goes out into space.

I have always found it difficult to believe that a star one million times the size of our Sun would produce light at the same velocity as our Sun. If there is a difference say of 1,000 times faster than light from our Sun how would that affect our vision of the Galaxy? Are there stars we cannot see from our Solar System because light density near Earth precludes their light or because the initial velocity is slow enough for it to dim out completely to our eyes due to the distance? This suggests that there could be solar systems quite close to us that we are unaware of because we do not see the light. They might see us but we could not see them.

The orbital velocity of the planets may include a factor of the density of light at the planets position as well as the effects of the Sun's gravitational pull. Sir Isaac Newton's Universal Law of gravitation recites that the gravitational strength diminishes inversely with the square of an objects distance from the source. Normally we would think that any two bodies in space would each exert an attraction for each other. Common sense suggests that a large massive planet would be farther from the Sun than a less massive planet. This is not the case in this Solar System. The fact is that the Sun's gravitation effect on the planets is to propel the closest planets in orbits at a higher velocity than those farther away without any apparent regard for the mass of the planet. This prompted me to observe previously that we could swap the positions of many of the planets and the planet in the new location would assume the orbital velocity of the planet that was relocated. If the light of the Sun plays a part in the orbital velocity of a planet then it is conceivable that the planets bulk could contribute to the effect, but I doubt it as to a planets revolution around the Sun and reserve judgment concerning the planets rotation that is likely effected.

In my projection above I used the square of the orbital velocity to deduct from the proposed 186624 speed of light at the minimum solar radius. Why squared? I am not really sure except to note that the conversion for the speed of light at Earth's orbit worked when the orbital velocity was squared. Also many other calculations I made required that the orbital velocity of the planet be squared.

This discussion requires some thought be given to the effects of a light source that is sending us light at speeds greater than our locale can accept. If I am correct there is every reason to expect that there are Stars 1000 times or 100,000 times as massive as our Sun that are sending us light faster than our 186,281 miles per second. What would we see of that light, possibly from a distant Galaxy, here on Earth? We would measure it at our speed here at Earth but what effect would the reduction in speed have on this light? A common effect that Astronomers see in the light coming to us from space is that it is red shifted. This view of light being red shifted is likened to the Doppler effect of sound changing as a train or other noisy object approaches us then passes us. There is a change in the quality of the sound so you can tell the object is going away from you. The red shifting is supposed to denote that the light source is going away from us. This means that the Universe would be expanding. Some light sources are blue shifted interpreted to mean they are not running away from us.

I think we have misinterpreted what the light is telling us. Light coming to us from light sources that travel faster than 186,281 miles per second are compressed causing the spectrum to show the red shifting. Light coming here at close to, or closer to, 186,281 miles per second may be blue shifted, but in any event not red shifted. How about light from a source that is from a small star traveling much slower than ours? Would we see it at all?

To elaborate on this it is necessary that I explain a little more. As light passes the Earth is speeds up going to Jupiter. From Jupiter or any other solar body it is reflected back to us at a faster speed than when it left Earth. This is automatic due to the distance traveled and light is in constant expansion as it travels and what causes it to travel. The reflected light returning to Earth is traveling faster and is less dense so it gives up the extra speed and is compressed to our density level bringing the speed to our level in Earths environment. Also we do not get all of the expanded light.

To make a test for this last explanation I would be very curious to know if when astronomers take the spectrum of reflected light from Jupiter or Saturn if there is any sign of a red shift in that reflected light. Saturn is 886,737,800 miles from the sun and therefore potentially 793,776,360 miles from Earth at our mean distance. Light from the sun travels to Saturn then back to us for a total of about 1,680,514,160 miles until we see it. I contend that the light from Saturn could be traveling at about 186,605 miles per second when it reaches Earth except it is compressed on the way back at each point wherein the local pressure is greater and gives up its greater speed as part of the compression so when it gets to the vicinity of Earth we measure it as 186,281 miles per second. This short distance and not too great of a speed difference may not show a red shift on testing. If it does we have potential proof.

If we consider a point in space one sun radius distant from the sun's equatorial surface, or 432,000 miles, and calculate the probable speed of light at that point we would take our Solar System radius of 31,830,914 miles divided by 432,000 and get 73,682.67. Deducting that from 186,624 gives us 112,941.3 which is the speed of light at that distance. It is curious that the square root of 73,682.67 is 271.44. The suns circumference of 2,714,342.4 divided by 271.44 equals 10,000 (or specifically 9999.787). This is where I found the multiplier for my proposed speed of light calculation. 432,000 divided by 10,000 equals the 432. The 432 squared equal the 186,624. There is no way for me to know if the same method would work for another Star to give us a hypothetical speed of its light. **As far as I know this is the first and only explanation of why the speed of light is measured at 186,281 miles per second.**

Chapter Fourteen
The Oblate Sun

I'm still working on the mathematics of why the planets rotate as they do but I had all of this other stuff on my mind so I will share it. As discussed in chapter thirteen the sun produces light which is not fixed at 186,281 miles a second. Our sun is a very large object made up of stuff that burns by atomic power providing light and heat and radiation and magnetic forces and solar winds and possibly some more things about which we have not yet learned. What ever I would venture to say on the subject was gleaned from the hard work of the professionals, meaning academics of all kinds.

The first and most obvious question is why the sun, or any star, keeps on going and going and going. The smart money says it is a self-sustaining generator/fusion mechanism. This conclusion assumes we can have billions of stars with no companions because the star is self-made and self-sustaining. I could never buy into that view. What do we know of in life that is independently self-sustaining? We know that our star, the sun, had company when it was formed. We are the living proof of that. To finish an argument for which no proof is available I will simply offer the intuitional conclusion that ALL stars have companions, either planets or a binary companion or both. Further more it is not possible to have a star any other way. The companions are required to keep the machine functioning just like they were required for the creation of the star in the first place. The great search for other planetary systems will succeed for the obvious reason that planets are everywhere. Your view may differ.

I want to tackle something that has intrigued me from the moment I first read about it. You may not have heard about it but our sun demonstrates "differential rotation"- not for me but for everybody else. There may certainly be some degree of differential rotation involving the Sun but I think the visual effect is due to the variance in the speed of light leaving the sun on its way to Earth. I will display my thoughts on this issue in another chapter.

I may have mentioned in my repetitive way of overlapping ideas that there had been a dispute among the professionals as to whether the sun was oblate or not. You may take note that just about every planet is oblate. That simply means that the planets belt line is a little larger than the top to bottom measurement due to the planets rotation which may be aided by the tug of satellites. Earth is a relatively solid body that is not turning very fast but it is oblate. This was felt to be important because Dr. Einstein suggested that the orbit of the planet Mercury was not uniform due to the gravitational effects of the Sun warping space. Mercury did have an unusual progression that they felt could be explained by the sun being oblate or if the sun was not oblate then by Dr. Einstein's theory. My guess is that everybody involved took a good look at the sun and they found no oblateness in it, Dr. Einstein's suggestion won out.

When I first read about this dispute I was curious as to how you go about deciding if the sun was oblate or not.

First of all we note that the sun is no where near as dense as the Earth (by their measurements) meaning that the sun is made of lighter stuff which hints to me that it would have more of a tendency to give way to a bulge in the middle. As I have no doubt shown elsewhere the Earths equator rotates at .2882 miles per second or you can say it takes the Earth 3.469 seconds to rotate one mile at the equator. They say the sun makes one full rotation in about 25 Earth days. 25 Earth days times 86,400 seconds provide that the sun takes about 2,160,000 seconds for one complete 360-degree rotation. The sun's circumference of 2,714,342.4 miles divided by the 2,160,000 seconds shows us the Sun rotates at about 1.256 miles per second and that is 4.36 times as fast as the Earth rotates.

To digress a moment I want you to note the curious relationship of Pi in these numbers. The 1.256 miles per second equatorial velocity of the Sun, times 2.5 = 3.14. This is a commonly used substitute for 3.1416. There are other curiosities but my notes on it are hiding right now.

Ok, we have a sun that has a radius (432,000/3963) 109 times that of Earth. A sun that rotates about 4.36 times as fast as the Earth and there is no visible oblateness! Keep in mind that the oblateness of an object is believed to be due to the centrifugal force that is generated by the objects rotation. My view sees the oblateness as a product of both the spin of rotation and the pull of the satellites that surround the object and the sun has the most objects pulling on it. I offer the note that both Mercury and Venus are depicted in the data as having a zero factor of equatorial oblateness and they have a zero population of satellites.

However I will confess that they lack a significant rotation velocity while not conceding that as controlling. When you take your average gas balls like Jupiter, and especially Saturn, we see oblateness that is greater than the Earths. With Earth reported as .0034. Jupiter as .062 and Saturn as .096 based upon the information I obtained from the "Technology Laboratories booklet of Space travel". Here we have a slight conflict. Jupiter's equator rotates faster than Saturn's. Recall that larger planets rotate faster than smaller planets. Saturn is less dense so it has more gas and with more gas you are likely to bulge more – so Saturn is more oblate for its size than either Jupiter or the Earth. You can guess why I went through all of these comparisons, but if not it's OK.

The sun has all of the basic ingredients that the planets have but there is no apparent oblateness. Something about Ducks comes to mind, but suffice to say that I believe that the sun is as ablate as the rest of us and that we do not see it because of the peculiar qualities of light that I spoke of in the last chapter. When you display the light from a common battery operated flashlight on to an object you are seeing the light reflected from the object. I have a keen edge on the obvious. The Sunlight on a planet or other object is no different when it comes to seeing reflected light. We can see the edge of the moon, more or less, because we are seeing reflected light. The Sun does not work that way because stars create their own light.

The speed of light as I propose is not fixed. It leaves the sun surface at a fairly fixed velocity but we do not see all of the light from all of the sun simultaneously because some points on the sun are farther from us than are other points. The top and the bottom are farther from us than the equator. The equator is accentuated due to the equatorial bulge. This in turn is due to the sun being oblate! The sun is rotating counter clockwise at the same time the Earth is revolving around the sun counter clockwise. Every part of the sun's surface is emitting light no matter where in space that particular spot is pointing.

I will try to make a rough sketch of what I mean but for now I will explain my view as well as I can. When we look at the sun we see the sun's hemisphere and that is the same image that the Earth presents to the sun except that Earths entire hemisphere is visible. Light is created on the entire sun's surface. The sun with a diameter of 864,000 miles presents a surface that is 432,000 miles more distant from us at the poles then it is the from the area at the sun's equator. The light from the poles and from the edges must travel 432,000 miles before the light from the equator engages it. This light from the fringe area of the sun is less dense than the light from the equator but it is traveling faster. I will offer that the initial speed of light for practical purposes is about 112,941 miles per second at blastoff. Light from every mile between the fringes of the sun working towards the sun's center must travel a little farther than its neighbor to reach the high point on the sun's equator.

The most distant point from us is equal to the suns radius of 432,000. Light starting from that point, if it was static at 112,941 miles per second, would take something less than about 3.825 seconds to reach the area of the sun's equator. Admittedly this is some pure speculation due to a lack of data as to the degree that the sun is oblate. The natural progression of the suns light velocity means that it would be traveling faster and faster becoming less and less dense on its way to the suns equator. This will be seen as pretty far out by astronomers and other academics involved with the space sciences. My reward would be just to have them read it enough so they can refute these suggestions.

The bottom line here is that the sun's edge and most all fringe areas will never be visible to us here on Earth. I have seen those wonderful pictures taken of the sun when eclipsed by the moon with large curls of fire that appear to disclose the sun's edges. When we look at the sun we never see the true edges because the sunlight that would normally disclose the edges is absorbed into the main stream of light giving us a picture of a ball reduced from the real image. No oblateness is visible even though it is there. This is not an easy claim to make. I am personally shocked that any theory of mine could make such unbelievable contentions.

The sketch that follows is meant to emphasize the area referred to as the fringes and distinguish them from the part of the equator closest to us here on Earth.

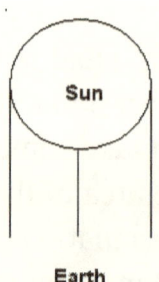

As you see here, light from the sun comes to us straight from the extremes and also from the center. The sun is so huge compared to the Earth that most of the light from the edges never becomes visible in a specific way that would disclose details. We have all seen pictures of a total eclipse of the sun by the moon. The moon, when in eclipse position, is only about 239,239 miles closer to the sun than the Earth and only 2130 miles in diameter yet it almost completely obscures the main stream light from the sun, and that, to me, lends support to my view of the limited source of the light that we see from the sun. Whether or not the sun is oblate as I contend and whether or not the sun's oblateness is a sufficient cause for the behavior of Mercury is of little importance to me. It was the debate between the academics that drew my attention to the issue. I had made my determination of the speed and the nature of Sunlight well over 20 years ago.

If I am completely in error with these views then that will be good news for everybody. Life can continue with the knowledge that all is well and the complications I suggest need not be of concern to any one. If I am right then some new appraisals will be called for all around.

The original source of the solar motion, I am satisfied, was the ever-increasing gravitational pull from the central parts of the Galaxy. There is no reason to assume that the original motivation ever ended. We are still captive and still influenced by that attraction. Once initiated the machine continues to function. Why does it hold its own? Sunlight may be the supplemental power source that sustains the motion of the planets and their satellites. I am looking into it.

I am aware that the sun's image will be reduced when it reaches the Moon.

Chapter Fifteen
Sun's Differential Rotation

From the previous discussion of the Sun's light effects in Chapter Thirteen and Fourteen there would normally follow other effects that result from these strange doings. My interest here is focused on an item that Astronomers refer to as "differential rotation" of the Suns surface. This is an outgrowth of observations of the Sun while Sun Spots were traveling across the Suns surface. The Sun rotates counterclockwise so we see the surface of the Sun turning toward us from left to right. It was noted that Sun Spots spread over the Sun's surface developed a pattern with the Spots in the vicinity of the equator moving faster than the other Spots farther out towards the Sun's poles. The rough sketch below assumes that the Sun Spots all start in a line on the left

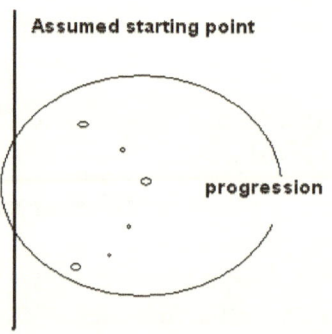

rim and there after move across the face of the sun with the equator area moving faster. It is believed that there are ribbons or speed zones causing the spreading of the Sun Spots. This effect may be due to other factors that I have tried to work out. You already know that I am not mathematically gifted so this would not be easy for me. I will be using essentially the same arguments about the speed of light that I offered before except that in this case, in an effort to be more precise, an attempt will be made to recite some numerical support.

I have explained in the last chapter that sunlight starting from various parts of the Sun's surface must travel different distances before all of the Suns light meets at some point in space from which it seems that all of the light travels together. The situation is quite complex in action but quite simple in the execution. The simplistic end result is due to the merging of the Suns light that obscures the less dense light and absorbs all the light into the densest light sources. We see a snapshot of the suns image from various points on the suns surface. The light from the equatorial area being the slowest and the densest, with the equator rotating, shows the Suns equator as appearing to be moving faster than the other surfaces. The snapshot from the areas close and closer to the poles contains light that left the Suns surface minutes earlier and show a picture of a gradual shifting of the Sun Spots which in reality have not changed their relative positions as to each other.

The idea of differential rotation that assumes different layers on the Suns surface moving at different speeds, and the example used to demonstrate the conclusion, fail in my mind because if that was true the only time you would see sun spots in a line is when they were at the leading edge of the Sun's sphere where about all the light is traveling at one speed towards us.

Remember it is all the light we see here on Earth that I am speaking about and not ALL of the light leaving the Sun in our direction. If you could see the Sun spots lined up at the sun's edge that would require that they formed there at that spot because if there really was differential rotation they would not hardly ever be in a vertical alignment by random.

We can try to dig a little deeper into this effect by dividing the sun into sections that roughly conform to the hypothetical speed lanes believed to exits in differential rotation. My divisions are just picked for a uniform depiction and are not intended to signify any observed speed lanes. It is an example of what I think is happening. Looking at the rough sketch on the right side I want to represent the sunlight heading off to the Earth from different points on the Suns surface.

Sun divided into light speed lanes.

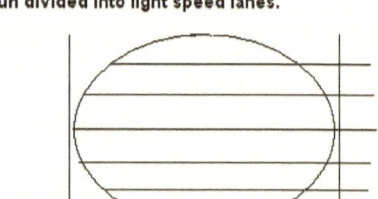

The upper most point, as with all of the other points, will provide light with a specific velocity. Each starting point of light will have a unique speed and a unique density by the time it gets beyond the Suns surface. I will assume a point at one sun radius from the Sun for the merging of all the light from all sources. That point in space will be a vertical line that for this discussion I will place at 432,000 miles distant from the sun's equator. For all of the light to merge there some of the light must travel farther than other light with the closest point being the sun's equator. My view is that light starts slower than currently thought and that it speeds up while it travels. My view also contends that the light will get less dense (weaker) as it travels.

I must reiterate this concept in more detail because I find that I get lost just in the description of what I claim happens. From the remote tip of the polar area of the sun, that point most distant from the sun's equator, there will be approximately 432000 miles straight out to the point where the equator ends (not providing for the sun's oblateness). I am not sure what the speed of light is as it blasts off from that point and we may never find a way to measure it. I am satisfied that it may be about 112,941 miles per second. I will conclude that it will go faster and faster as it races to the point where our hypothetical vertical line is at 432000 miles from the sun's equator. That light source should be going at about 149,827 miles per second when it reaches our hypothetical line in space. The light that blasted off from the equator would be starting at about 112,941 miles per second. The light from each point between the poles and the equator would blast off at 112,941 miles per second and take relatively shorter times to reach our hypothetical vertical line 432,000 miles distant from the sun. For instance, if we take a point half way between the pole and the equator the start velocity is still the same but the distance from the start point to the hypothetical line would be less so the light would be traveling slower than the light from the poles but faster than the light from the equator. Wow, that is a mouthful. So what's happening?

The light from the pole must travel 864,000 miles to reach our line in space and that will take something less than 7.65 seconds. The light from the equator will take about 3.825 seconds to reach the hypothetical line in space and that is half as long.

The areas between the poles and the equator will each take more time that will be proportional to the distance of each starting point. All of the potential light will take, based on averages, about 500 seconds, or about 8 minutes to reach the Earth. We see the fastest light that emanated from the equator as the controlling light giving us the illusion that the equator is rotating faster than the upper and lower parts of the sun. It is important to envision the difference in the shape of the Suns surface due to the curvature of it.

The book "Contemporary Astronomy" estimates that the rate of observed differential rotation causes the suns area at 40 degrees to rotate in about 28 days as opposed to the Suns rotation at the equator in about 25 days. That is 3 days slower. Let's go back to the velocities of the suns surface at the equator that we know is about 1,256 miles per second. 28 days times 86400 seconds means that a sun day at 40 degrees is 2,419,200 seconds and that gives us 1.12 miles per second for the hypothetical surface velocity of the sun at 40 degrees. This will show the 40-degree area to be traveling about half as fast (1.256 / 1.121= 1.121) as the area at the equator. All other areas of the sun's surface will fall in line for a proportional and illusionary view of the sun demonstrating "differential rotation".

I think we get a snapshot of the suns surface at our hypothetical point in space which is the equivalent of a still picture of the suns surface depicting what ever there is to see, including sunspots, that appear to be moving at different speeds across the face of the sun because the light bringing us the picture is an illusion. We do see apparent differences in the speed of Sun Spots on the face of the Sun but they are not moving at different speeds as we see it. Having said that there is still the possibility of some degree of differential rotation.

There are so many mysterious challenges out there, and here at home, that I sometimes wish that I were born smarter and more open brained so I could understand them. In law school the important issues in the cases were not always apparent. Some said the test material was designed to learn how many issues the student could decipher in the question. If you fail to note in your analysis that one important issue and/or that secondary issue you would fail the test. Possibly you had a good grasp of the law on the subject but for reasons not apparent you were unable to apply your legal skills to the question at hand. Later quite possibly the points were clear. My approach to Astronomy, a subject about which I have had no training, was to keep looking at the facts and see what might be hidden there. This can be done with many challenges. When you come to what you believe to be valid results you must stand by them even though they run askew of the majority view. There is no way or means by which I can be dogmatic about my views but I trust them never the less.

In conclusion I do not believe that the sun exhibits differential rotation to the degree the academics see it. I think that what the astronomers see is an illusion caused by the variable speed of light and the fact that the sun is oblate as was discussed in chapter fourteen.

James J. Wood, Sr.

So far I have been unable to suggest a test that would serve to prove or disprove my contentions as to the speed of light and its various effects. Some smart academic will do it.

Chapter Sixteen
Let's Go to Mass
And a little on Surface Gravity

I have never had an adequate grasp of Isaac Newton's Laws of Universal gravitation. Possibly my aversion to long miniscule measurements is a factor but more likely I am not quite up to it. There is no limit to the people that use the formula very easily. Is it likely that the more miniscule the yardstick is the more accurate the result will be? If the planets were very close to each other in size then a more definitive measure would be very necessary to get some comparisons. The second reason is likely that because the purpose is to find the amount of 'force' involved precision is mandated. I felt it would be worthwhile to go on with my miles mass workup in spite of my shortcomings. I obtained what I think are more recent calculations of the planets mass in grams from TheFreeDictionary.com on the Internet for comparison.

Solar Object	Mass $-x10^{23}$	Mass /Earth	Mass Miles	Mass Miles /EarthMM
Sun	19891000.	332,982	Sr.	328,079
Mercury	3.302	.05527	5,413.82	.0558
Venus	48.685	.815	79,073	.815
Earth	59.736	1	97022	1

Mars	**6.4185**	**.1074**	**10,319**	**.10635**
Jupiter	**18990.**	**317.90**	**30,378,202.7**	**313.11**
Saturn	**5684.69**	**5.162**	**9,118,897.5**	**94**
Uranus	**868.32**	**14.53**	**1,372,202.7**	**14.14**
Neptune	**1024.3**	**17.14**	**1,663,774.**	**16.88**

As noted in the earlier chapters the miles mass result always produces a smaller mass ratio for other planet/Earth comparisons. Mercury and Venus do not count because I did not make a unique determination of miles mass for those planets. I do not contend in any way that my miles mass is more indicative of the facts. The difference, if understood, could lead to some worthwhile development. You may recall the difficulty I had with Uranus when trying to get a useful miles mass result using the average of all the satellites figures. The only obvious trend here is that the more weight massive the published data is for the object then the more pronounced the difference is compared to my miles mass figures.

The sun is significantly less massive by my method than as published. 332,982-328,079 provides a difference of 4,903. If we take the relative mass ratios of all the planets listed above we only get about 446. The overall difference between the Sun and the Earth by this approach suggests that difference between the Earth miles mass and the published mass for the Sun/Earth is about enough to account for all of the mass in the Solar System. If so the published mass for the sun is exaggerated or my miles mass for the Earth is diminished.

Reducing the exponential numbers provided for the System based on Newton's weight in kilograms by –x23 we can compare the weight mass ratios to the miles mass ratios.

The sun's weight mass in kg is 19891000. That divided by Jupiter's weight mass of 18990 = 1,047.446.
The sun's weight mass/Earths are 332982. That divided by Jupiter's of 317.9 = 1,047.446. the weight ratios equate.

My Sun's MM/Earths MM = 328079 and that divided by Jupiter's 313.11 = 1,047.8. That is almost exactly the same.

The ratio of the values remains essentially the same. I suppose that may mean that my formula is doing a reasonable comparison of mass with some peculiar difference that shows a reduced mass when compared to Isaac Newton's formula. Let's try Saturn's results

The sun's weight mass in kg is 19891000. That divided by Saturn's weight mass of 5684.6 = 3499.1.
My sun's miles mass/Earths MM = 328079 and that divided by Saturn's 94 = 3490.2. The difference may be in rounding.
The result is very close but not identical. It appears that the ratios resulting from my method may be consistent for all planets.
I also want to compare the Sun full MM of 31,830,914,183, which is the Sn or SS# number.

James J. Wood, Sr.

Mercury: MM = 5,413.82
Sun mass Kg at 19,891,000 / 3.302 = 6,023,924.89
Sun MM divided by Mercury MM of 5,413.82= 5,879,566.4
Mercury MM/Earth MM of .0558 = 5,879,552.

Venus;   MM = 79,073
Sun mass Kg at 19,891,000 / 48.685 = 408,565.3
Sun MM divided by my Venus MM of 79,073 = 402,551.
Venus MM/Earth MM of .815 = 402,551.
No miles mass was independently projected by me as of yet.

Earth: MM = 97,022
Sun mass Kg at 19,891,000 / 59.736 = 332,981.78
Sun Sn MM divided by Earth MM of 97,022 = 328,079.
328,079 / Earth MM = 3.381. (A trial item)

Mars: MM = 10,319
Sun mass Kg at 19,891,000 / 6.4185 = 3,099,010.67
Sun MM divided by my Mars MM of 10,319 = 3,084,689.8
Mars MM/Earth MM of .10635 = 3,084,898.9

Jupiter: MM = 30,378,202.7
Sun mass Kg at 19,891,000 / 18,990 = 1,047.66
Sun MM divided by Jupiter MM of 30,378,202.7 = 1,047.82
Jupiter MM/Earth MM of 313.11 = 1,047.8

Saturn: MM = 9,118,897.5
Sun mass Kg at 19,891,000 / 5,684.6 = 3,499.1
Sun MM divided by Saturn MM of 9,118,897.5 = 3,490.65
Saturn MM/Earth MM of 94 = 3,490.2

Uranus: MM = 1,372,202.7
Sun mass Kg at 19,891,000 / 868.32 = 22,907.45
Sun MM divided by Uranus MM of 1,372,202.7 = 23,196.94
Uranus MM/ Earth MM **of** 14.14 = 23,202.2

Neptune: MM = 1,663,774.
Sun mass Kg at 19,891,000 / 1,024.3 = 19,419.11
Sun MM divided by Neptune MM of 1,663,774 = 19,131.75
Neptune MM/Earth MM of 16.88 = 19,435.95

The mass miles radius for the sun works so well for the mean orbital velocities of all the planets that I feel it is a reliable ratio. We find above that it holds true for the miles mass comparisons as well. All of the miles mass figures are based on averages of the satellites distances product. Any average is likely to give the results a somewhat vague quality.

My figures for Mercury can only be suggestive because I have not as yet perfected a means of getting MM for planets with no satellites. Mercury and Venus may be over estimated for Kg mass. No independent mass miles were worked out for Venus.

My figures for Jupiter suggest that Jupiter is right on target for published mass and it works for the miles mass ratio. Jupiter has practically no equatorial tilt.

My figures for Saturn suggest Saturn's Kg mass may be under estimated. I had an idea that the tilt of the planet would affect its apparent mass- and I still think it does. I thought my miles mass ratio would show some difference with the ratio provided by the Kg mass. The lack of a difference leaves the door open for further workup.

My figures for Uranus suggest that Uranus mass in Kg is underestimated. Uranus with a tilt of more than 97 degrees is a good example for my theory concerning apparent mass.

In this instance the retrograde rotation of the planet and its major satellites exaggerates the effect we would find with only tilt to contend with. Providing proof that Uranus mass is underestimated requires a means to calculate a planets probable mass by some method that does not rely on the orbital characteristics of the satellites. I want to pursue this avenue even though I may very well fail in the effort. Note that Isaac Newton's Law compares the distance between orbiting bodies to work out the formula. You may recall the difficulty I had searching for the best mass miles ratio for Uranus. The larger retrograde satellites produced a much smaller potential miles mass result when compared to the small satellites in counter clockwise orbits. My view is that the orbital velocity and the orbits mean radius is affected by the favorable combination of the planet and the satellites revolving in the same direction. When Uranus was flipped over the large satellites went with it so that everything was then in retrograde rotation. The small satellites very likely were acquired after the fact and assumed normal counterclockwise orbits.

The published data shows Uranus to have an equatorial diameter of 32,187 miles. Neptune is shown as having an equatorial diameter of 30,447. These diameters would produce a volume comparison wherein Uranus volume is 1.18147 times that of Neptune. I am well aware that volume does not directly equate to mass but if all other factors were equal and these planets were made of the same stuff then Uranus should be about 1.184147 times as massive as Neptune. The potential density of each planet is not a factor in this discussion because density is determined by dividing the volume of a planet by its mass in grams. If the initial determination of the mass in grams is faulty then the density conclusion is equally faulty. My hypothetical assumes for comparisons that each planets density would be about equal subject to size considerations wherein there is the prospect that larger spheres would tend to compact the matter to a greater degree by gravitational forces.

The comparison of Uranus and Neptune brings up another important question about the formation of the Solar System. Why do they find such a variance in the comparative densities of the planets? The most obvious answer is that the substance from which the planets were formed was not uniformly distributed when the System was in the making. So we must conclude that there was heavier stuff in some places and that the quantity of the stuff was greater in some places. It seems that there is no common denominator for the making of planets. At one time I tried to envision a planet much like an onion with layers of matter getting successively less dense as each layer was closer to the surface. The idea was to discover a uniform growth pattern that would apply to all objects thinking that gravity would make such a situation likely because it uniformly applies to all. I did not find what I was looking for but the idea is still enticing. One potential problem is that the stuff of creation was not uniform so that adhesion and compression would not be uniform. I will not pursue that dead end any further- I mention it only so you will see one wasted effort.

The density of the planets and satellites being so different is very curious. If all things in a vacuum fall at the same speeds what is happening to make one object denser than another? If gravity was the cause for the objects collection of substance in the first place and gravity was the cause for the stuff to compress into a sphere what is the force that compresses some objects more than others? My thought is that the size of an object will equate to its density even though the size may not have a bearing on the mass of the object. I keep looking for uniformity. I find total randomness illogical and quite unlikely. If Newton's Laws of Gravity are universal then there must be some action-taking place that we have not discovered. Towards the end of this chapter I will be discussing my effort at surface gravity using my miles mass. Normally,$= 9.79$ m/s^2 a s I believe it is done, the volume of the object, probably in Kilometers or smaller, is divided by the mass of the object in grams, or some thing like that. The formula is not all that simple considering the data required making it work.

I have learned that all of the data that the academics so carefully accumulated is very inter dependent. If a conclusion regarding the mean orbital radius of an object is a little off it will affect almost all of the conclusions about the objects relationship to other objects. Any conclusion relative to mass will affect our view of the overall Solar System. I have been extremely impressed with the detail and overall apparent accuracy of the academics observations.

The comparisons for Neptune suggest that Neptune's Kg mass is probably correct. The >28-degree inclination of the equator suggests that there is an acceptable tilt that will not affect the apparent mass of the planet. Possibly up to 30 degree tilt is not meaningful otherwise there may be some underestimated mass in Kg for Neptune.

I was optimistic that there would be a way to derive an objects surface gravity using miles mass even though I knew it would be completely foreign to the established method.

SURFACE GRAVITY

Surface Gravity, I found, was not so easy to come by. I tried to repeat the calculations of others wherein the one formula looks something like this:

$$g_e = \frac{g\,M_e}{Re^2} = \frac{(6.673 \times 10^{-11})\times(5.97 \times 10^{24}}{(6.378 \times 10^6)^2}$$

These formulas were obtained on the Internet at: http/farside.ph.utexas.edu/%7Erfitzp/teaching/301/lectures/node152.html,

Or, I'm told, using Newton's law of gravitation it is known that any spherical body of mass M and radius R possesses a surface gravity g given in this formula:

$$\frac{g}{g_e} = \frac{M/M_e}{(R/R_e)^2}$$

I never tried to run out the formula although it should be easy enough for others to do it. My interest was to see if there may be a way to do it with the miles mass figures. When surface gravity is published in textbooks it is usually done with reference to Earth.

They show Earth as "1" and all other planets as they relate to Earth. The speed of a falling object to Earth is 32.16 feet per second, per second. This means an additional 32.16 feet velocity for each additional second. Using that as our Earth's surface gravity value we get the results in the following layout:

Mercury	.38	FPS	12.22	Radius	1,515.5
Venus.	.91		29.26		3,759.5
Moon	.17		5.46		1,080.0
Earth	1.		32.16		3,963.3
Mars	.38		12.22		2,111.5
Jupiter	2.5		80.40		44,678.7
Saturn	1.07		34.41		37,284.0

Uranus	.88	28.30	16,246.5
Neptune	1.14	36.66	15,379.6
Sun	28.	900.48	432,000.

What you have above is an interpretation of the published charts to convert numbers dependent on Earth ratios to some numbers that stand alone as to their individual performance. I relate the first column to Earth's 32.16 feet per second and portray the result in the second column in actual feet footage.

My miles mass efforts were used to attempt a formula that would relate to surface gravity to see if it could be done. My best effort was to use the planets miles mass times 5280 feet divided by the planet radius squared to get my surface gravity. Let's try it out.

	Miles Mass **Mile Feet** **Result** **Radiussquared** **Result**
Mercury	**5,413.82 * 5280= 28,584,969 / 1,515.5²= 12.44**
Venus	**79,072.93* 5280= 417,505,070 / 3,759.5²= 29.12**
Moon	**((5.46*(1080²))/5280)=1,207 miles mass.** **(1,207*5280)= 6,372,960/1,166,400 = 5.46.** **I used surface gravity to find miles mass first.** **Earth 90722*.0123 = 1,193.4.**
Earth	**97,022* 5280= 512,276,160 / 3,963.3²= 32.61**
Mars	**10,319 * 5280= 54,484,320 / 2,111.5²= 12.22**

Jupiter $30,378,021*5280=160,395,950,080/44,678.7^2=$
80.35

Saturn $9,118,897*5280=48,147,778,80 \ /37,284,0^2= 34.63$

Uranus $1,372,202.7*5280=7,245,230,256/16,246.5^2=27.45$

Neptune $1,663,774.0*5280=8,784,726,720/15,379.6^2=37.13$

Sun $Sn\#*5280=168,067,232,584,480/432,000^2= \ 900.56$

I did the Moon in reverse to find miles mass for the Moon working backwards from the published surface gravity. This was done because I had never worked out a miles mass for the Moon on my own.

Considering the simplicity of the formula the results are quite reassuring. The results, when compared to the published data are quite close. I will defer to the professionals and assume their figures are most accurate. It is really curious that the results for Uranus are farthest from the published data. My guess would anticipate that Uranus by my method would show up as more massive than theirs- not so.

I want to try an experiment using the volume of Uranus divided by Neptune as worked out earlier at 1.18147 and multiply that by Neptune's miles mass of 1,663,774.4. We will get 1,965,699,.54, * 5208 = 10,378,893,573. That divided by 263,948,762 gives us 39.32 as a revised surface gravity. This is just slightly more than Neptune and within reason although it is all guess work for now.

Going back to Chapter Ten, I added up all of the eleven small satellites and the average was 1,396,381.2 miles mass for Uranus. The four retrograde satellites averaged at 1,366,660.05. This difference made my overall average less than perfect. In the above layout I did an alternate Uranus projection with the slightly larger miles mass with little actual difference.

There's much to be done on the issue of an objects potential mass based on posture. I feel it may be beyond the purpose of this book. I think my speculations are correct but unproven.

SOME UNPRODUCTIVE EFFORTS:

Earth's position number of .000789/7926.6, Earth's diameter, = .00000009953. = E
Moon's position number of .002874/2160, Moon's diameter, = .00000133078. = M

M/E = 13,370642. The square root is 36.5658885 or X.
Earth diameter of 7,926.6 / X = 2,167.75 or almost the Moon's diameter.

The Moon's mean orbital radius is about 239,239 miles If we rerun the position number for the Moon at 400,000 miles orbit radius and rerun the formula above the Moon's diameter becomes 3,187.127 miles. So farther appears larger. If we rerun the Moon's position number at 200,000 miles orbit radius and the do the above formula the Moon's diameter becomes 1895.07, which is smaller than the normal size in the proper position. Every time we move the Moon we will have a different orbital velocity for the new orbit that will in turn potentially affect the miles mass for the planet. The affect will be to provide a larger, or smaller, miles mass for the Earth based on the Moon's position getting farther or closer. I think that all current measurements of mass, in terms of relative weight, rely on this kind of orbital criteria.

If a planet or satellite is not in harmony with the system then I think that the satellite's orbit may be retarded potentially providing a false result in the determination of the planet's weight. This needs more support than I have shown thus far. I tried the above method on other planet **satellites with** results that were not productive.

However, my previous formula using the position number works very well for the Moon's statistics as well. ((2160 *3.1416)/.002874)=2,361,119 seconds in Moons mean orbit,/86,400 = 27.32 days.

Chapter Seventeen
A solution to Rotation

When I started my Solar System search for some simple way to decipher the basis for the rotation of the planets I did not think there would be any necessity for me to learn more about the formulas of others. Isaac Newton's work was foreign to me and difficult for me to comprehend. The physics of the system should be left alone if I wanted to try something new using miles as a measure instead of kilometers. I did not like to work with kilometers and I did not feel the measure contained the same synergistic relationship to time that miles did as a measure. I am now at Chapter Seventeen and I would like to reconsider some of the stuff that got me here. I am aware that the weight of the planets was worked out long ago. I was never looking for a means to weigh the planets or the satellites. I wanted to look for something different. My views on the nature and the speed of light originated from an entirely different mindset and had nothing to do with my efforts on the Solar System except as they may bear on the rotation of the planets. Newton's work explains how it is that the planets maintain position and orbital movement. No point in going there again.

After playing with the numbers for a long time and running many Basic computer programs that I had written it became clear that a new approach was required if I hoped to decipher the basis for the rotation of the planets. I did not want to use Earth related numbers that projected data that provided data uniquely related to Earth. I also became aware that the days and years of a planets orbit around the sun were destined to be useless when seeking planetary rotation criteria. Orbital velocity was equivalent to a planets period and provided a more detail method of comparison. I was unwillingly confronted with a history of kilometers and kilograms which I disliked working with so right from the start I used miles for my yardstick. This was fortuitous because I learned quickly that there was a working relationship between miles and seconds. So I searched for a way to proceed. We are told that the mass, weight, of an object in space is the base line for the affects of gravity. In its most simplified description it means that the heavier the object the more gravitational pull it will have meaning the greater attraction it will have for other objects. Any effort to weigh the objects would be foolish in light of this work having been done long ago. Some other method was required.

It was well established, I thought, that if the Earth mass was sufficient to propel the Moon at a specific velocity there was a proven relationship between the Earths mass, the orbital radius in miles distance between the moon and the Earth and the speed at which the Moon traveled in it's orbit. I thought that was well established for the sun/planet relationships as well. These are the traditional factors in standard calculations with the addition of Newton's constant. A different approach, I thought, may provide a simple means of getting a mass ratio that would be useful without regard to the potential weight of the objects. This will contain some repetition which I feel is required because of my contention that the weight mass of a planet has nothing to do with the planets period of revolution around the sun as will be shown by the calculations repeated here.

I used the mean orbital radius as equivalent to semi major axis. They say that the mean orbital radius of the Moon is 239,239 miles for a diameter of 478,478 miles. The diameter * 3.1416 provides a mean orbital circumference of 1,503,186.48 miles. We are told that it takes the Moon 27.32 earth days to make one trip around its orbit. The orbital circumference of 1,503,186.48 miles divided by 27.32 earth days equals 55,021.467 miles a day. 55,021.467 miles a day divided by 86,400 seconds a day equals an orbital velocity of .63682253 miles per second. If the Earth is capable of orbiting the Moon at .63682253 miles per second what is the ratio, in miles, of the Earths mass for rotating a moon at 1 mile a second?

I obtained the result by the square of the orbital velocity of .63682253, miles per second, which equals .40554. This number times the orbital radius of 239,239 miles equals 97,022 as the radius of the mass of the Earth expressed in miles. There remained to be decided if the result is the mass miles for only the Earth or is it the value of the Earth Moon combination. I saved that inquiry for later. My reason for using a velocity of 1 as the measure of the planets force is that the number 1 cannot be squared. This provides a uniform measure for comparing the planets vital statistics. Beyond a velocity of 1 mile per second the velocity will be a fraction of 1 and continue to diminish until some other solar force interferes.

It seemed to me that this method could also be used to measure the Sun's gravitational force in miles. Using the earth as my tool I checked and note that we are told the Earth's mean orbital radius distance from the Sun is about 92,961,440 miles. This times 6.2832 provides for an orbit circumference of 584,095,319.808 miles. It takes us 365.25 days to make 1 trip around the Sun. 584,095,319.808 divided by 365.25 equals 1,559,165.83 miles a day. 1,559,165.83 divided by 86400 seconds equals 18.5 miles per second. Using the same method as above I square 18.5 for 342.25 times 92,961,440 getting 31,816,052,840 as the mean radius of an orbit around the Sun with a velocity of 1 mile per second.

To recap, the Sun's mass in miles radius is 31,816,052,840 = minimum Solar System radius. The Sun's mass in miles diameter is 63,632,105,680. The Sun's mass miles circumference is 199,906,623,204, which I rounded up to 200,000,000,000 miles.

This meant that this Solar System has a minimum diameter of 63,632,105,680 miles. I have been using the probable circumference of the Solar System, at 1 mpsv, as 200,000,000,000 miles. I referred to this as my SS#. There were many interesting relationships to be found in this conclusion. The system proved to be so mechanical that I could take the seconds in a year for any planet and divide it into the SS# and I would get the orbital velocity, in miles per second squared, for that body. This method showed that the Sun's mass miles radius of 31,816,052,840 / Earths 97,022 showed that the Sun is 327,926.17 times as massive as the Earth. One textbook describes the Sun as 330,000 times as massive as the Earth and another source makes the Sun about 332,948 times as massive. This concerned me because it again forced the consideration that my ratios might be faulty. I assumed that the academics results were secured using Newton's laws. The difference was significant and was to be discussed later. It is possible that I was getting a distorted measure of the planets and the satellites.

The next logical question is to learn how the Sun relates to this SS# in miles. The Sun with a diameter of 864000 miles provides for a circumference of 2,714,342.4 miles. The SS# at 200,000,000,000 / 2,714,342.4 = 73,682.67 its square root is 271.44. This, presented another way showed $2,714,342.4 * 271.44^2 = 199,991,862,070$, miles which is the SS# before rounding.

Thereafter, without getting into the Asteroids, I went through the Solar System proving that the orbital velocities for all the planets in the system conformed to the formula with out regard to the weight or mass of the object. This was a surprise to me because I expected that the acceleration of the planets in their orbits to be due to the combined effects of the mass of the planet and the mass of the sun. Probably I do not understand the formula that I think says it.

$$(Fgrav = (G* Mplanet * Msun)/R^2)$$

My efforts at seeking the source of the planets rotations disclosed that mass or weight appeared to have nothing to do with a planets rotation. That was also apparent when determining the planets period. Size was everything with small planets turning slower and large planets turning faster. There were three apparent ingredients that might be affecting the size of a planet and that appeared to be the sun's gravity, the sun's rotation and Sunlight itself. Knowing what we are told about Sunlight having no mass and therefore weightless it appeared that was not likely the cause for the rotations. I was satisfied that the sun's gravity was contributing a rolling ball effect to the inner surface of the planet and that, if there was a force pushing the planet around its orbit that might be the answer. Note that if the sun's gravity was holding on to the inner face of the planet, and some other force was causing the rotation that would be more likely contribute to a planet rotating clockwise. That is not the case. Those planets or satellites that rotate clockwise are the freaks of the system. The next issue was what could push the planet around its orbit if we decide that the respective masses of the involved bodies appear to have nothing to do with the orbital velocity of the planet? The culprit appeared to be Sunlight with the force being the uniform expansion of sunlight as it travels through the Solar System. This conclusion then prompts another issue. If the density of Sunlight gets weaker and weaker as it travels through the system how could there be a uniform effect on the planets that are farther and farther away from the source? To pursue this thought requires the contemplation that the rotation of the planets would experience a diminished affect of the suns efforts as we go farther from the sun. This latter suggestion does not appear to be a factor and tends to diminish the prospect of the sun's contribution.

I think we must go back to the origin of the Solar System. Before there was a sun and sunlight there were differentiated gases and substances spread wantonly through out a large area. This matter had no form and very likely due to its lack of form the gravity of the galaxy simply carried the substances along as a large body of diffuse matter. After some point in time, possibly after this substance had made many trips around the galaxy, this material began to assume a loose shape collecting sufficient mass to draw the attention of the central galaxy, which brought to bear its gravitational attraction. I did not find a specific reference to the rotation of the galaxy but the pictures of galaxies suggest to me that they are rotating clockwise. My approach, as with most things, is simple. The clockwise motion and attraction of the galaxy slowly caused the solar substance to form a disk that itself began to rotate counter clockwise. This is somewhat like one wheels edge turning another wheels edge. This I believe was due to the central attraction of the galaxy having a greater attraction for those parts of the solar substance that was closer than it did for the more distant parts. This effort caused a disk to form comprising all of the material with greater concentration of matter at the center of the disk. Before there was ignition, before there was light, the system shaped itself into many ringlets that were themselves revolving around the central core in a counter clockwise manner like the main body of the assembled matter. For this reason I contend that the uniform counterclockwise revolution and rotation of the objects in the Solar System was due to this creation process. This is no doubt old stuff to the academics but I recite my thoughts so the process is clear as to how I get where I am going. This may make one wonder why one gigantic planet didn't form at the center of this collection as all of this stuff was collecting towards the center. My thought is that throughout the collection of this matter, rivers or circles of matter were forming rings from the center out to the very edge of this disk of stuff. These rivers were in motion around the forming central mass continuously collecting more and more stuff from which to form something of their own. Why would the disorganized matter form into these rivers separating themselves from each other? The issue remains concerning the formation of planetoid spheres taking place before or after the ignition of the sun. Most likely there would be a compromise with the spheres partially created but still growing by their mass attracting more of the substances that were still diffused. I think the main event was triggered by the ignition of the sun when the sun achieved enough preliminary mass to start the fusion reaction. This power produced extensive gravitational waves and light that sustained the motion of the Solar System and the final formation of planets and moons that in turn sustain the sun.

I have been very interested in the sun's rotation as a factor in the revolution of the planets as well as the rotation of the planets. The sun is said to make one full 360-degree rotation in about 25 Earth days. 25 times 86,400 seconds gives us 2,160,000 seconds in one sun day. With a diameter of 864,000 miles the circumference of the sun is about 2,714,342.4 miles. The sun's circumference of 2,714,342 miles divided by 2,160,000 means that the sun rotates 1.25664 miles per second at the equator. If the sun is oblate as I contend and if we do not know the degree of that oblateness then this result will be slightly off of perfect- but I use it anyway.

Sunlight does not appear to turn with the sun's rotation. Each shot of light goes straight out into space unless obstructed or bent by the interaction of other objects or, conceivably by the sun's gravity that could cause some movement in the direction the sun rotates... This can be pictured in slow motion as if the sun shoots out a beam of light constantly as it turns moving the light beam infinitesimally to the left. If this is true then we would have the combined effects of Sunlight and gravitational waves working on the planets, this means that the right front quarter of the planets hemisphere is getting a progressive concentration of light pressure. I thought that might be the cause for the planets rotation. I have not resolved that issue yet but I did explain the right front pressure on the planet and the left back quarter getting none in a previous chapter.

The disk of the sun's gravitational field may turn with the sun, as the sun turns. This would not be like a phonograph record with ever increasing speed out to the edge of the disk because we know that the force of gravity decreases inversely with the square of the distance. I have provided many examples of the diminished orbital velocity of the planets as you get farther and farther from the sun. I have also shown how that force is diminished, just in the distance involved with the diameter of the planet, when I arrived at my significant numbers for the planets and the satellites. You will recall that if we consider the different speed lanes like tracks in space we found that the inner surface of an object is in a faster track than the surface more distant from the orbited object.

Let's digress a moment and reconsider the circumference of Earths orbit around the sun. With a radius of 92,916,440 miles the circumference will be about 584,095,319.8 miles. If the Earth simply rolled around this circumferential highway it could rotate as many times as the Earths circumference divides into the orbit circumference. That would be about 23,455.759 rotations or "days". This means that there is some slippage in the system. When we divide the 23,455.759 ratio by the actual days, of 365.25, we get 64.218. I was optimistic that there must be some known number that will decipher 64.218 into something useful. We have plenty of numbers to play with but surprisingly that number turned out to be the orbital velocity in miles per second. 64.218/18.5 = 3.47. The Earth rotates 1 mile at the equator in 3.47 seconds. Or 1/3.47 = .28818 which I had previously rounded up to .2882 miles per second rotation for planet Earth at the equator. This looks perfect. Because the orbit and the planets circumference work we can use the radius for each item instead.

I think this is worth trying on the other planets. The formula would read:
1/((orbit radius/planet radius)/actual days)/orbit velocity mps)= svmpm.

Planet Orbit Radius	Planet Radius	Roller Real days	Days	Result Orbit vmps	surface velocity
Mercury					

(35,977,324/ 1515.5)=23,740 / 87.96= 269.9/ 29.75 = 9.07.

    This planet does not rotate normally. I must try to give it a normal rotation. 1/9.07 = .1102 which is my projected equatorial surface velocity for normal rotations. You may check Chapter thirteen for the previous workup for Mercury, Venus.

Venus
(67,232,236/3,759.5)=17,883.3 / 224.6= 79.62 / 21.76= 3.66.
This planet is in reverse. 1/3.66 = .2732 mps surface velocity is my projection when normal rotations resume.

Earth
(92,961,440/3,963.3)=23,455.5 / 365.25= 64.22 / 18.5= 3.47.
We know the data. 1/3.47 minutes equals .2882 mpsesv.

Mars
(141,617,060/2,111.5)=67,069.4 / 669..62=100.1 / 14.99= 6.68.
Mars works as observed.1/6.68 = .1497 mpsesv.

Jupiter
(483,822,040/44,678.5)=10,828.9 / 10,620 / 1019 / 8.11 = .1256.
1/.1256 = 7.962 my miles per second equatorial velocity.

Saturn
(886,695,015/37,284.)=23,782.2 / 25,247.76 / .942 / 5.99 = .1572.
1/.1572 = 6.36 miles per second equatorial velocity.

Uranus
(1,783,953,320/16,246.5)=109,805 / 47,601.18/2.307 /4.22= .5466
1/.5466 = 1.829 mps equatorial velocity.
Some doubt in the published data.

Neptune
(2,794,363,106/15,289.5)=182,763 / 91,324.52 /2.001 / 3.38=.592
1/.592 = 1.689 mps equatorial velocity.

I am pleased to note that the method does work for all of the planets

The calculations for the approximate days or rotations recited in the formula above are very simple. I am using the academics observed rotations for each of the planets. I start by converting the mean orbital radius into an orbit circumference in miles. The circumference is divided by the orbital velocity in miles providing the seconds in a full orbit. The total seconds of one complete orbit are then divided by the seconds in one complete rotation. The result should be fairly accurate for the total rotations in one complete revolution

I was very pleased with the above formula but it was not the final result I was looking for. This method requires more data about the planet than the formula I was seeking. The planets size is essential. Some number representing the force is essential. The system is so precise numerically that the next resolution of numbers in my effort to narrow down the original formula demonstrated that the relation ship of the orbit radius divided by the planet radius, such as 23,455.5 for the Earth, when divided by the mps for Earth orbit of 18.5 mps gave me my original significant number for the earth, or 1267.8, or $1/1267 = .000789$. If you reverse the process you get the miles per second orbital velocities for each planet. That was an interesting discovery for me but not the solution I was seeking.

The suggestion is that the orbital velocity of the planet and something relating to the planets size might work if I kept at it. In previous chapters I showed that the significant number for a planet when multiplied by the planets actual year's days gave us the equatorial mps velocity. That required the insertion of actual days and I would like to eliminate that requirement. The significant number was previously worked out for the planets and the satellites by the following means which I repeat here as a reminder. Recall SS# = 31,830,914,184.

(SS#/(92961440+3963))= 18.5039257773^2 **Farthest is slowest**
(SS#/ 92961440)   = 18.50432018963^2 **Center is actual velocity**
(SS#/(92961440-3963)) = 18.504714627182^2 **Closest is the fastest.**

I deduct slowest from the fastest to get a position number.

18.504714627182
<u>18.503925777300</u>
.000788849882

1/.000789 = 1267.349 is easier to work with and is resurrected here.

Why does it work? The potential Earth rotations in a full orbit around the sun are 23,455.5. I showed that dividing that number by the actual Earth days led to a number that when divided by the orbital velocity gave us the the apparent equatorial surface velocity for the planet. If I had published satellite observed rotations I would have devoted a chapter to the formula working the satellites.

This number, has some unique qualities in the system for Earth and I suspect that its peers involved with the other planets do also. Note that 23,455.5 * .000789 = 18.5, which is Earths orbital velocity in miles per second.

Still working with Earth's number of 23,455.5 I continue to search. Old friends pop up now and then like, (23,455.5*3.1416) = 73,687.8. The square root of which is 271.45. Recall that the sun's circumference is 2,714,342 / 10,000 is 271.43. I keep getting sidetracked with the interplay of numbers. ((73,687.8^2/864,000)/2) = 3,142.3.

Very close to Pi. I then want to know why that is so but there simply is not enough time left for me to pursue all of the things that I hit upon that I find interesting. Also ((864,000*Pi)*73,687.8) = my SS#. Enough number playing.

We now know we can get this unique result by applying (1/((Or/Pr)/V))). 92,961,440/3963.3=23,455.5/18.5=1,267.86

Newton's view was that the sun's gravity was the reason for the planets being held in position to revolve around the sun. I question that he ever speculated as to what caused the planets move in their orbits or to rotate as they do. Had Newton applied his genius to that issue I would have lost out on many years of fun and frustration and would instead be writing about the Bible, the Pyramids of Egypt or some more fanciful subject?

Having tried to consider all of the ingredients that could contribute to a planets rotation I have manipulated many numbers many times and I am still searching. There is a way to shorten the effort by using the significant number acquired for each planet with the actual rotations for that planet. Rather than recalculate the Sn for each planet above we can use the formula above that will provide it. We still require the actual day's rotations. This will be Days or rotations * ((Orbit velocity / (Orbit radius / Planet radius)) = rotation of equator in miles per second.

Planet	Days	mps
Mercury	87.96 * (29.75 / (Or/Pr)) =	.1102.
	Or 87.96 * .001251279 =	.1101.
Venus	224.6 * (21.76 / (Or/Pr)) =	.2732.
	Or 224.6 * .00121684 =	.2733.
Earth	365.25* (18.50 / (Or/Pr)) =	.2881.
	Or 365.25 * .000789 =	.2882.
Mars	669.62* (14.99/ (Or/Pr)) =	.1496.
	Or 669.62 * .00022347 =	.1496.
Jupiter	10,620.00*(8.11/ (Or/Pr)) =	7.954.
	Or 10620.00 * .00074409 =	7.9
Saturn	25,247.76* (5.99/ (Or/Pr)) =	6.359.
	Or 25247.76 * .00025191 =	6.36
Uranus	47,601.18* (4.22/ (Or/Pr)) =	1.952.
	Or 47601.18 * .00004104 =	1.953
Neptune	91,324.52* (3.37/ (Or/Pr)) =	1.684.
	Or 91324.52 * .00001823 =	1.665

The complicating differences of length of orbit and velocity of orbit have made it difficult for me to find a "constant" that would apply equally for each planet. The above represents the best effort by any one that I am aware of in this world.

James J. Wood, Sr.

Asteroid Ceres at 747.3 * .00001345 should rotate at .010052 mps at the equator if my data is correct.

It is possible to search for answers that do not exist. When I consider the accuracy of my formulas for getting the surface velocity of these objects I am impressed. It was a long time coming but I feel it was time well spent even if there are some that think there is no point in knowing these things. If time permits there are a lot of numbers whose origins should be investigated to learn why there is the interplay that we see.

All in all, I think the most mysterious thing of all about the Solar System is the way the numbers work, as if by design. I am not a seriously religious person but it looks like some entity, a person or a God played a part in the Systems design.

Chapter Eighteen
Conclusion

There are some unusual ideas expressed in this book. Some of the work will seem redundant to the reader and I will agree it is so. Much of the numerical work was done to set the stage for more in depth comparisons of the distances between objects and the size of objects. This was done in a potential search for the gravity waves I feel are in control. There will be some back lash on the speed of light and I welcome the dissention for no other reason than it would mean somebody read my book.

My first goal was to demonstrate that Miles makes the best measure of this Solar System. I think I have shown that clearly. This was not all by accident. I think some smart people before our time worked out a system of measures to do, in an even more simplified way, what I have done. Going back over my notes to see if there were any loose ends I came across an "Appendix" I copied from some textbook that contained "Basic Constants" of space mathematics. Pi is shown as 3.1415926536. It may have been a step towards more accuracy if I had used that constant rather than the 3.1416 that I used, but I will stay with what has been done. More important, these constants, I'm told are those adopted as "IAU (1976) system of astronomical constants" at the General Assembly of the International Astronomical 'Society', the word after Astronomical was missing so I speculate it was 'Society'. The important point here is that I have previously confessed   having no special training in any field related to the content of this book and no special training in mathematics.

Imagine my surprise to find listed as a constant "e". It said e = 2.7182818285. There was something familiar about that number. To learn a little more I went to the Internet, MSN Encarta, and found the information, e (mathematics). We are told that e, in mathematics, is a very important constant comparable only to Pi in a wide variety of situations. You can read it for yourself if you want to learn more about "e". As I have mentioned many times the circumference of the sun when expressed in miles is 2,714,342.4. Note the inclusion of e. My minimum estimated circumference of the Solar System is 200,000,000,000 miles. That divided by the sun's circumference of 2,714,342.4 equals 73,682.67 and that has a square root of 271.44. That is almost exactly e times 100. No doubt there are always going to be some that will argue this is just a coincidence of numbers. I think that this is by design and not by accident. I have shown plenty of examples that miles are the measure of the Solar system because they work so well with time. Days and seconds flow in meaningful ways that are not possible with meters. My approach to surface gravity using mass miles is an example of the simplification of measures even if they do not fit the published data exactly. More work is required.

If the more distant planets reflected light comes back to us with a red shift that may lend support to my theory about light being compressed here on Earth when emanating from other sources. If not red shifted that will not necessarily mean the theory is wrong because we will be talking about reflected light and not original light sources. The discussion of the sun's differential rotation was suggested by my theory of light and the unusual behavior of the light we see due to source and speed differences. Right or wrong I hope you found it interesting to contemplate. I sure did when it originally occurred to me. The issue of an oblate sun has more involved than merely my views about the nature of light. There is also the fact that all the planets show some oblateness and they have a lot less tugging at them than the sun does.

Scholars here have measured the speed of light at Earth many times. As far as I know there has never been any person in this world that has offered a formula or an explanation for light traveling at about 186,281 miles a second. I have provided both in this book and that is a first. Why would light be stagnant at one speed start to finish?

The miles mass approach was, as I commented on earlier, required for providing me a means to compare planets and satellites in a way that differed from the customary manner provided by Newton's work. I think it worked very well in the limited way I applied it. My goal was to eventually get to a division of the results to define the respective mass of the main planet and the satellite. I didn't get that far because I simply ran out of steam. My 76 years are beginning to weigh on me a little and I wanted to get these thoughts published. After so many years in the making of notes and calculations I feared for its loss. If my destiny permits I will finish this exploration later and pursue any loose ends.

There should be little doubt that the more obvious dysfunctional objects in the System were not likely that way when the System was originally formed. Mercury does not rotate in a proper manner. I tried to suggest a probable rotation for mercury based on the performance of the entire clan. If there is something going on between Mercury and the sun to hold Mercury in its weird rotation then it may be something like the Moon in Earth's orbit, which we are told is based on the distribution of its mass. As far as I know the dispute about whether or not the sun was oblate had nothing to do with Mercury rotating or not. It dealt with revolutions.

Venus is in reverse, a situation that is only possible by Venus being tipped over as I suggested. The planet is extremely hot and that is due to the reverse rotation, some compression that probably occurred and some contribution of the cloud cover. The present conclusion for the greenhouse effect is based on the cloud cover that I see as only a part of the problem. I did speculate that Venus was at one time beyond Earths orbit and was a large object when seen from Earth due to the reflective clouds, the size and the proximity to Earth. NASA's people have advised they are confident that Venus had loads of water at one time. Check out the article below.

(Times Mirror Company Los Angeles Times, March 25, 1993, Thursday Home edition. "Venus once had abundant water, Space Probe indicates". Byline: Mark A. Stein, Times Staff Writer. Venus, the arid and intensely hot "twin planet" of Earth, once was temperate and covered by perhaps 75 feet of water --. National Aeronautics and Space Administration Sciences said Wednesday.

I saw this data much later as confirmation of my view. It was not a cause for my view. So where did all that water go? Earth, I believe, got a great deal of it when Venus traveled from where it was to where it is now. I have not been able to come up with the cause although the obvious answer is that some massive object moved things around and now we do not know where it went. I feel it is beyond debate that Venus was formed just like the rest of us and was rolled over by something. I am sure an academic skilled in the ways of physics and Newton's Laws could give us a pretty good idea of how much mass would be required to tip Venus over and how swift the object must be moving to leave it that way. That would be true for Uranus as well. We will not live long enough to see it but Venus will eventually start rotating normally and much of the cloud cover will turn to water again. The planet will then cool to normal temperatures as it once enjoyed and will be a great place to visit.

The Earth, as I see it, was originally a smaller planet than what we have now. The large ocean trenches and the drifting of the continents are, I believe, residuals of the Earth's previous expansion. The Earth has seen catastrophes of which the Biblical flood is only one. When discussing Noah's flood, wherein the mountains were covered, my favorite question was "where did all that water go?" A flood of that magnitude does not simply go underground. The survivors would be unaware of the noise and upheaval of mounds and mountains because the water would muffle the sounds. I think the Earth expanded at that time in keeping with the extensive mass of water added to the planets mass.

Many learned people, in different fields of learning, have brought to light serious changes in the geography of the Earth and the rise and fall of landmasses. These events left their mark. The horrendous inundation of Earth at various times made us the water soaked planet we are today and the planet Venus contributed its share.

When you think about the Earth, and compare it to its neighbors, you should wonder how we came to have such a large Moon. Mercury and Venus have none. Mars has two rocks that are not even spheres so they don't count. There seems little likelihood that our Moon could have formed with us during the creation period. It formed some place most likely in this Solar system but not next to us. When I get to speculate about things the facts are usually fairly clear and it is my explanations that are speculative. We have Mercury almost not rotating and Venus going backwards. We have the Earth that may have dealt with some size changes and some relocation. The Moon, when put in this picture, should be expected to be someplace other than where it was originally. My guess is that it was the first object from the sun. The dark gray volcanic looking surface suggests a very hot environment to me. All kinds of trash heading for the sun would find it a welcome target towards the end of the System's creation. It is unimportant but that is where I would put it. For the Moon to get from a spot closer to the Sun to the orbit around the Earth is not such a great leap of conjecture as might be thought. There is a delicate balance of objects with the effects of gravity being a slow and weak action is insufficient to insist on stability in the face of a celestial confrontation. In my view anything out there can be moved.

So, how about Mars? This planet seems normal enough. If I were to landscape the Solar System I would put Mars after Mercury. My plan would be the Moon to be first followed by Mercury, and then followed by Mars. Then Earth followed by Venus. The missing planet between Mars and Jupiter, which I name Atlan, was destroyed when everything was being reorganized. These speculations are not important to the books other content. It is just fun to contemplate them.

Jupiter is magnificent. Why is it where it is? Because of its size and the diminishing size of the planets after Jupiter we might think it is midway in the Solar System. Not so. I have offered what I contend is the minimal radius of the Solar System, If we divide Jupiter's mean orbit radius by my minimal Solar radius we get 483,822,040 miles divided by Sn# which means that Jupiter is only .0152 of the distance into the full Solar radius. All of the large planets are fairly close to the sun when considered in the light of the Solar System's probable radius. Assuming Pluto has a mean orbital radius of 3,674,648,900 miles and that it is the farthest member we know of in this System it measures in at only .1154 of the distance into the full Solar radius. I wonder at why the mass of the System appears to be concentrated so close to the center of the System?

The size of the Solar System seems to have been a mystery of its own. As I finished up this last chapter I went to the Internet a few times to see what was new for me.

Times Mirror Company, Los Angeles Times, May 27, 1993.

> Evidence of Solar System's edge found.
> Byline: Associated Press
> The twin Voyager space probes have found the first evidence of the existence of the true edge of the Solar System, where solar wind hits interstellar space, NASA announced Wednesday.

The article discusses the data providing the news but never offers a statement to tell us what the size of the Solar System is. This form of scientific discovery is frequently lacking a punch line. Great. You have discovered the Solar System's probable edge. Where is it and how far is it from the sun? No comment. Well, I calculate it is probably something more than a radius of 31,830,914,183 miles to the edge.

At one time tried a run on the planets by dividing the diameter of the planets individually into the circumference of the sun. This effort was prompted by my discovery that the circumference of the sun divided by the diameter of the Earth gave me the orbital velocity of the Earth squared. It did not work for all of the other planets but it did work for Mercury, Earth and Saturn. If the procedure was followed in reverse you could reorganize the System based on the various mean orbital velocities disclosed. I decided this was not a worthwhile pursuit and went no farther with it. It is of interest that Mercury and Saturn worked out.

The distance separating objects is used to measure the mass of objects, in part. A planet will appear more or less massive depending on how far its satellites are from it in their orbits because more mass will equate to more attraction. When I read about Newton's work I noted he did not finalize his calculations until some other person worked out the weight of the Earth. I have no idea how he did that. The volume would be easy enough to calculate, but the weight stumps me. We are in a weightless environment within the Solar System. We weigh something in relationship to the Earth and we would weigh something in relation to whatever planet we were on. Possibly that weighty person worked it out based on how things fall to Earth. No matter because he did it and Newton used it to advantage. To me this means that Newton's formula could not disclose the weights of objects in the System until he had an example from which to find a common denominator and from that derive a "constant". When I read his formula for determining the force of gravity between objects and mass I always wonder if moving some of the objects farther away from the orbited object will cause it to look more massive as a result of the formula. If the distance is important then the distance will affect the result assuming all the other factors remain the same. I never got to completely finish my own effort using the miles mass but I did satisfy myself that the alleged mass of any object was going to show as the product of the formula and not necessarily be factual. When using my miles mass approach I could not isolate a pattern for different miles mass for a planet due to the objects distance in orbit around it. There were differences, to be sure, but there was no pattern showing more distant objects supported a more massive planet. The only exception to this last statement is the satellites of Uranus wherein the normally orbiting satellites supported a slightly greater mass. The one fully consistent finding I can call uniform is that the miles mass method always produces a less mass result when compared to the weight mass method. Much of the difference was when I compared the miles mass of other planets to the Earth. The why of it requires more work? I'm taking one last look before going to the publisher but expect nothing.

I firmly believe that Uranus is under rated because it is on its side and does not get the extra push for the satellites that the more upright planets enjoy. As part of this idea I spent a lot of time years ago attempting to derive a relationship between the volume of a planet and the reported mass and I was unsuccessful. I thought my problem was that I was unable to come up with a progressive density to be considered along with the volume. Along the way I came to realize that the density of the planets was very diverse and had no apparent connection to a planets size other than that the formula used considered the planets volume. In spite of my inability to demonstrate mathematically that Uranus is more massive than previous calculations show I am convinced that it is. Any thing over a 30-degree tilt from the vertical will, I believe, cause a faulty mass result when compared to normal planets. We can conclude that Uranus was not born that way. With the four major satellites revolving retrograde it appears logical that they were there at the beginning and turned over with Uranus. The balance of the satellites in normal counter clockwise revolutions probably joined Uranus after the calamity that caused Uranus to roll over. It was not striking but we did see a difference produced by the miles mass method between the counter clockwise moons and the clockwise moons with the counterclockwise moons producing more miles mass for Uranus.

While on the subjects of rotating satellites I am again brought back to the cause of revolution of planets and satellites. I was of the view that Newton's law accounted for the revolutions of the planets. In an effort to get a little education, and not to over do it, I went back to the Internet, (Copied by be about March 10, 2005)

http://www-spof.gsfc.nasa.gov/stargaze/Sgravity.htm (2/11/2005)
They provide, for me, the best and most understandable outline on Newton's Laws and how to do the calculations. I printed out 4 of the 5 pages. You do it.
On page two about 2/3rds of the way down the page they recite:

> "(Please note: gravity is not what gives the moon its velocity. Whatever velocity the moon has was probably acquired when it was created. But gravity prevents the Moon from running away, and confines it to some orbit.)"

I should have found this source sooner. It seems that neither the basis for the planets revolutions or for their rotations was fully discovered before. I should have devoted more space to the issue, and would have, except that something I read implied that Newton's work explained the orbits and the forces working on the objects. When it comes to the sun and the planets I think we can conclude that it is the sun's source of gravity that provides for the orbits of the planets *even though* the mass of the planet does not appear to play any part whatever in the result. The velocities of the planets orbits, as I have shown, are position precise and fall perfectly in line with the proposed approximate Solar System's radius. I find no other uniform force working on all of the planets other than the sun's gravity so I conclude that the orbits of the planets, in terms of their orbital velocities, are due to the sun's gravity. This can be visualized if we think of the sun on a large plate with a circumference of about 200,000,000,000 miles. This plate represents the area of the suns primary gravity within which all major planetary systems perform. This plate is rotating at some speed consistent with the suns rotation and it sweeps objects along with it. The sun also emits sunlight and solar wind that contribute to the final result. As I explained earlier the initial rotation of the Solar System was imparted by the effect of the galaxies draw on the stuff of which the Solar System was composed. This action was continued by the means mentioned above. The forces involved are weak but very persistent and continuous.

The planet serves only as a position functioning object and does not appear to contribute to the arrangement. If the planets mass has naught to do with the radius of its orbit how do we use the orbit radius in a calculation to determine anything about the respective objects weight? I know this is some failure of mine but I keep getting confronted with it. I used the satellites orbital velocity to obtain a miles mass factor from Jupiter's satellites, as I did for the other planets that have satellites. These efforts gave me a useful miles mass for the planet that could thereafter be used for more comparisons to get some relative miles mass results. It makes sense to me to find the mass, weight or in miles, for a planet from a satellite but it does not follow that would provide the mass of the satellite. As I said, it is my thing to resolve. I will restate it for the last time. The revolution and the rotation of objects in this Solar System are the result of the sun's or planets gravity, the added result of the sun's or planets rotation and the expansion of sunlight and solar wind pressures.

I am satisfied that the main purpose of this book was accomplished even though I left some loose ends for some other smarter person to tie up for me. The reason that no one ever worked out a formula for the rotation of the planets is apparent now. I gave that issue a lot of attention and many times thought that I had the solution only to find it was not. From the formulas I offered in this book anyone could see that a "constant" such as that derived by Newton for calculations of gravity was not found. I did provide formulas for a number of things that can be used to double-check the results of academic observation of the planets and satellites. The rotation issue could only be expressed in surface velocity in minutes or seconds because each planet has a different size and a different orbit. Sir Isaac Newton's formulas require the mass of the objects, the distance between them, Pi, and his Constant to make them work. My method of getting the surface velocity of planets equatorial in miles per second requires only the planets actual days (rotations) and my significant number. I do not believe that method will work for non-spherical rocks tumbling through space. Hopefully I have set the stage for that lucky someone to find the best answer. My methods do work. They are not as simplified as I would have liked. I have no idea if any one else ever tackled the problem. When I told a good friend of mine of my pursuit, he said, "Why would anyone care?"

I have not used all of my various disorganized notes and calculator tapes. The things that appeared more useful or more interesting appealed to me and I hoped they would appeal to my audience. There are so many items we do not know about our Solar System that I wonder at times why the academics jumped to the Universe so fast. No doubt that effort could also provide answers to things here but it seems to me a round about way to do it. Of course, the Universe is of extreme importance and it is a lot of fun to speculate about the way it works. Newton devised his Universal Theory of Gravitation right here on Earth and it turned out to be a big window on the entire Universe. The Solar System may very well have as much potential information about the workings of the Universe as it does about the Solar System itself.

There are some important items in this book relating to the speed and nature of light. I hope some effort is made to test the theory. The expansion or not of the Universe may be explained by the spectrographic red shift being due to compression of light as I suggest. Also if there is such a thing as a degree of red shift I would like to see that observed from the largest stars to see if they produce a greater red shift consistent with light traveling faster from larger stars.

Still going back through some of my unused note I came across some efforts of mine that now look to me as if I was trying to find the origins of Newton's formula. Some curious things happen with numbers. One example was Newton's mass for the sun and the Earth less the 10^{25} showing the following:

The sun's mass, 198,910,000, times the Earths mass of 597.42, times .0000000667 equals 7,926.2. As you should note from the past numbers that is the diameter of the Earth at the Earths equator. How could that be? This suggested to me at the time that there might be a relationship between size and weight mass. In Chapter Sixteen I showed some mass figures as -10^{23} to simplify the calculations. I do not know now which source showed the values as 10^{25} but that is what my notes contain. The ratio is correct either way but the decimal is moved. This curious result is not duplicated with the other planets. I mention these dead ends in case a reader wishes to pursue them.

A situation came up often wherein a divisor on 10 or 10 xs gave meaningful results. One example is Earth's orbital velocity squared at (18.505 * 18.505) = 342.435. 342.435 divided by 3.1416 = 109. and 109 divided by 31.416 = 3.47, which is the Earth's equatorial velocity in seconds per mile? The number 109 came up a various times but I could not make use of it.

The best I have to offer thus far is in this book. The search for more simplified mathematics will be continued if fate favors me with the time and my attention is not distracted for some other or more interesting mystery more suited to my manner and method.

About the Author:

Jim was born in Philadelphia 76 years ago. A Catholic grammar school graduate and a public school dropout his marriage was the turning point in his life. Before getting married he had a short ride as a merchant seaman and two paid European tours by Uncle Sam; one voluntary and one not. After marriage and children Jim was encouraged to rethink his life and pursued the necessary catch-up to qualify as a law student in California. After four years of night school Jim was sworn in as a Lawyer in 1961 and was employed by a firm until 1964 when he opened his own office. Jim was a successful trial attorney specialized in administrative law until retirement in 1994. Jim has no formal training in Astronomy. He used the astronomers findings.

As far back as he can remember his interest was focused on the mysteries of life. He was an avid reader of everything he could find on such subjects. His hobby eventually focused on the Solar System. When family and work permitted he spent his free time on the numbers of the Solar System originally searching for the reason for the rotation of the planets and satellites, and a mathematical means to predict same, he eventually went into the fundamental workings of the Solar System as he saw it. This book is the culmination of most of Jim's ideas and notes created over about 30 Years